Study Guide to accompany

Human Physiology

Third Edition

RHOADES & PFLANZER

Florence C. Ricciuti

SAUNDERS COLLEGE PUBLISHING
Harcourt Brace College Publishers

Fort Worth Philadelphia San Diego New York Orlando Austin
San Antonio Toronto Montreal London Sydney Tokyo

Portions of this work were published in previous editions.

Printed in the United States of America.

Ricciuti; Study Guide to accompany Human Physiology, 3E. Rhoades & Pflanzer.

ISBN 0-03-015863-x

567 095 987654321

Preface

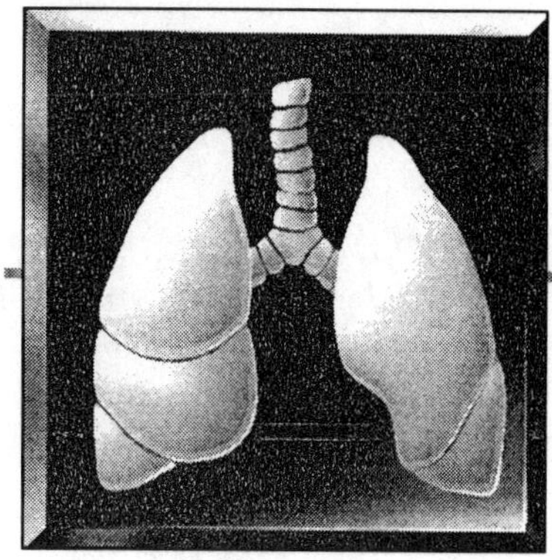

My objectives in writing this study guide were to:

- Summarize important concepts and facts
- Organize the content to facilitate student learning
- Provide a format for review
- Test student comprehension
- Develop student problem-solving ability

Each chapter in the study guide contains a:

1. List of learning objectives.
2. Detailed outline of the text chapter.
3. List of key terms and phrases.
4. Section to evaluate student learning.

The learning objectives, as well as the outline, stress the content to be mastered in a course of human physiology.

The expanded chapter outline follows the text chapter and contains additional headings and details. Major concepts and key terms are highlighted. The outline may be used in class to facilitate note taking and to help students review for examinations.

The skills understanding section is designed to:

1. Evaluate mastery and application of concepts.
2. Develop problem-solving ability.
3. Organize information into tables that may be used for review.

The exercise format is varied: Matching, Dichotomous, True or False, Sentence Completion, and Multiple Choice. Most chapters also contain tables to complete and text figures on which to identify structures.

Answers are provided for all exercises and most of the tables in the skills understanding section.

I welcome comments, criticisms, and suggestions for improvement, and would appreciate being notified of inaccurate statements or errors.

FCR
New Haven, CT
August 1995

To The Student

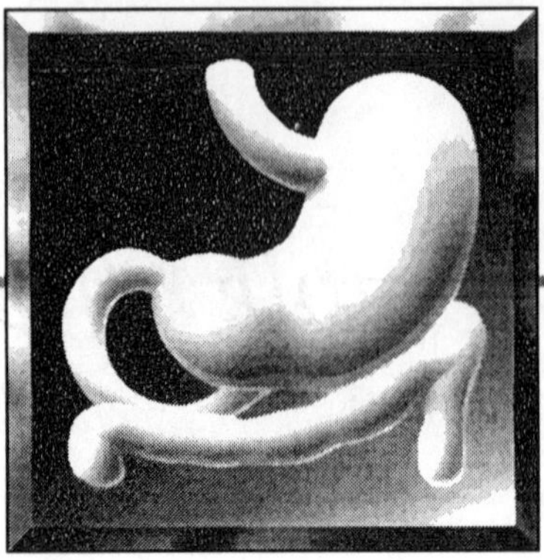

As you study each chapter in the text, keep the following points in mind:

1. Look over the text chapter outline and read the chapter summary to get an overview of the material before class. Also, skim the text chapter outline in the Study Guide.

2. Read the chapter in the text. Do not try to absorb or memorize every detail. The quick reading will help you become familiar with terminology, basic concepts, and text.

3. As you reread the text chapter, compare it with the expanded chapter outline in the Study Guide. During lecture, fill in details in the Study Guide outline.

4. When you feel confident with the material, complete the section on skills understanding. Answers are provided for the first five test formats (matching, dichotomous, true and false, sentence completion, and multiple choice), and most of the tables. Completion of the tables will help you organize material to prepare for an examination.

I hope this Study Guide is a useful tool for your study of human physiology. I welcome comments, criticisms, and suggestions for improvement.

FCR
New Haven, CT
August 1995

Table of Contents

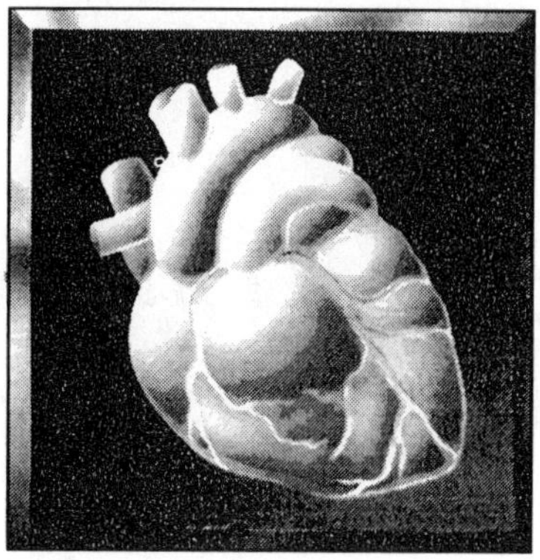

Chapter 1

The Science of Physiology

Chapter Learning Objectives

1. Trace the early history of physiology.
2. Identify early scientists and their contributions.
3. Describe why errors in interpretation of human physiology persisted into the Renaissance.
4. Explain how the early concepts of circulation were modified by simple observations and experiments.
5. Relate how our understanding of the lungs and their functions has changed.
6. Outline the steps in the scientific method and explain how it is applied.
7. Know the units of the metric system, how to interconvert between them, and how to express a quantity using scientific notation.
8. Distinguish deductive from inductive reasoning and provide an example of each.
9. Distinguish among the various types of cells.
10. Describe the role of homeostasis in human physiology.
11. Identify some of the quantitative measures used in physiology.
12. Explain the concepts involved in control systems in human physiology.

Expanded Chapter Outline

I. The History of Physiology

A. *Physiological knowledge in classical and medieval periods was steeped in dogma.*

1. **Aristotle** (384–322 B.C.) was one of the first to describe blood vessels as a system with the heart at the center and assert that blood flowed between arteries and veins.

2. **Herophilus** (335–280 B.C.), who is considered the first physiologist in the Western tradition by some, identified the brain as the seat of the intellect.

3. **Erasistratus** (310–240 B.C.) postulated the existence of tiny connections between veins and arteries, but he believed air was present in arteries rather than blood.

4. **Galen** (c. 130–201 A.D.) demonstrated that blood flowed through arteries and believed that different forms of pneuma, or "spirits," drove various bodily processes.

B. *The discoveries of the European Renaissance period revolutionized physiology.*

1. **Versalius** (1514–1564) wrote the first modern anatomy textbook, which was published in 1543. It contained many illustrations and corrected many of Galen's errors.

2. **William Harvey** (1578–1657) demonstrated experimentally the direction of blood flow in veins and the connection between arteries and veins.
   a. Harvey recognized that the heart is a pump and is not the place where blood and air meet.
   b. Harvey used quantitative methods to show that the heart pumps the same blood, which movs in a circular path in the body.

3. **Malpighi** (1628–1694) confirmed microscopically the existence of capillaries.

C. *Modern advances in physiology opened new frontiers.*

1. Modern understanding of air and its importance to the body can be traced to the work of 17th and 18th century physicists and chemists, such as Priestly, Lavoisier, Boyle, Borelli, and others.
   a. **Priestley** (1733–1804) demonstrated, using a candle and mouse, that a substance is removed from air during combustion and breathing.
   b. **Lavoisier** (1743–1794) identified the substance as oxygen, which he claimed sustains the body's innate heat.

2. The concept of the cell theory was articulated by Schleiden and Schwann in the 19th century.
   a. **Virchow** (1821–1902) first hypothesized that cells come from preexisting cells and not from non-living matter.
   b. **Pasteur** (1822–1895) experimentally confirmed Virchow's hypothesis.

3. **Bernard** (1813–1878) proposed the concept of homeostasis in the 19th century.

II. The Study of Physiology

A. The scientific method is a systematic pursuit of knowledge.

1. The **scientific method** involves posing questions and then searching for answers to those questions often by devising experiments and tests of reasons to validate the answers.
   a. In **deductive reasoning**, one proceeds from a general rule or premise to a specific conclusion.
   b. In **inductive reasoning**, one goes from a specific observation to a general conclusion, which can be destroyed by a single exception.

2. The scientific method consists of a series of steps broadly classified as: (1) observation; (2) hypothesis; (3) experiment; (4) conclusion; (5) retesting (sometimes); and (6) publication.
   a. The **hypothesis** tentatively explains the observations.
   b. The purpose of the **experiment** is to determine whether a cause-and-effect relationship exists between an observation and a variable (the hypothesized cause of the observation).
   c. Two sets of conditions are devised: (1) the **experimental** set, in which the variable is present and (2) the **control,** which is identical to the experimental set except the variable is omitted.
   d. The **conclusion** drawn from the results can be used to validate or reject the hypothesis.
   e. **Retesting** may be necessary to accept the hypothesis more confidently.
   f. Through **publication** new information is shared.

B. Measurement, computation, and graphic analysis provide a quantitative approach.

1. Quantitative methods are necessary for understanding bodily functions. The **metric system** is almost exclusively used for physiological measurements.

2. Quantities are most conveniently expressed exponentially.

3. A **ratio** compares two numbers or quantities by division. A **proportion** is a mathematical statement of the equality of two ratios.

4. A **graph** expresses a relationship between two or more quantities.
   a. The relationship may be direct or inverse or show another form.
   b. Through **interpolation** and **extrapolation,** predicted values may be placed on the graph.

C. Anatomical planes and positions provide spatial orientation of the body.

1. Imaginary planes are used to divide the body into parts.
   a. Three commonly used planes are **sagittal, transverse,** and **frontal.**
   b. Terms of position (e.g., medial, lateral, distal, proximal) are used to locate one structure relative to another.

III. General Organization of the Body

A. Human cells can differentiate and exhibit over 200 different modes of specialization.

1. Different groups of cells work together to maintain the life of the organism.

2. Each cell manifests the form and structure best suited to the function it performs in the body.

3. Living matter is made of protoplasm, which displays attributes of life.

B. Tissues are organized colonies of specialized cells.

1. One or more types of specialized cells become closely associated to form a **tissue.**

2. There are four classes of tissues: epithelial, connective, nervous, and muscle.
   a. **Epithelial tissue** consists of closely packed cells arranged in flat sheets and covering exposed body surfaces and

lining body cavities, tubes, and organs. It protects, absorbs, filters, and secretes substances.
b. **Connective tissue** is composed of cells embedded in a matrix that may contain cellular products, such as fibers, soluble proteins, and crystalline complexes. It supports, protects, binds, and partitions nearly all bodily components.
c. **Muscle tissue** consists of elongated cells that contain contractile fibers.
d. The primary components of **nervous tissue** are **neurons**, cells specialized for receiving and conducting electrical impulses, and **glial cells**, which provide neurons with structural and metabolic support.

C. Organs are integrated into systems.

1. An **organ** is defined as a group of tissues that have been combined for the performance of a specific function or series of related functions.

2. Organs and other structures that share in the performance of related tasks are grouped into **systems**.

D. Body fluids are divided in two compartments.

1. Bodily fluids are grouped into two major compartments: **intracellular fluid** and **extracellular fluid**.
   a. The compartments are separated by the cellular plasma membrane and differ chemically.
   b. Intracellular fluid is contained in subcellular compartments and has a relatively low concentration of sodium ions and a high concentration of potassium ions.
   c. The reverse is true for extracellular fluid.

2. There are three types of extracellular fluids that make up the internal environment of the body:
   a. Interstitial fluid
   b. Plasma
   c. Lymph

IV. Homeostasis: Control of the Internal Environment

A. A stable internal environment is essential for normal cell function and provides a degree of independence from external conditions.

1. Conditions in the internal environment must be maintained within narrow limits to allow cells to live and function.

2. Conditions such as oxygen tension, pH, osmotic pressure, temperature, and concentrations of metabolic substrates, hormones, and waste products are all closely regulated by the cooperative workings of tissues, organs, and systems.

B. Homeostasis maintains protein conformation.

1. Each protein has its own distinctive conformation that allows it to best carry out its function.

2. The shape and therefore the function are affected by pH, temperature, and ion concentrations.

C. Coordinated body activity requires a control system.

1. Control systems use mechanisms called sensors to detect conditions in the body and effectors to change conditions in the body.

2. Two general classes of control systems are active in the body:
   a. **Negative feedback** systems, which stabilize and work to restore normal values.
   b. **Positive feedback** systems, which destabilize.

3. **Sensors** determine the level of the variable with respect to a reference point, and **effectors** attempt to change the level of the parameter so that it returns to the desired level.

4. The nervous system, one of the major homeostatic control systems in the body, monitors temperature, arterial blood pressure, and concentration of hydrogen ions in extracellular fluids by thermoreceptors, baroreceptors, and chemoreceptors, respectively.

Key Terms

anatomical planes (sagittal, frontal, transverse)
cell theory
connective tissue
control system
cranial, caudal
deductive reasoning
direct relationship
effector
epithelial tissue
experimental design
extracellular fluid
extrapolation
homeostasis
hypothesis
inductive reasoning
inferior, superior
interpolation
interstitial fluid
intracellular fluid
inverse relationship
linear relationship
medial, lateral
muscle tissue
nervous tissue
observation
organ
positive feedback
proximal, distal
ratio
reference point
scientific method
scientific notation
sensor
tissue
negative feedback
variable (independent, dependent)
ventral, dorsal

Aristotle
Bernard
Galen
Harvey
Hooke
Malpighi
Pasteur
Priestley
Schleiden
Schwann

Testing Your Knowledge

I. Matching. Select the correct term from the Key Terms.

________ 1. proceeding from a general rule to a specific conclusion

________ 2. proceeding from a specific observation to a general rule

________ 3. determined correct pattern of circulation of blood in the body

________ 4. developed concept of homeostasis

________ 5. position of stomach relative to spinal cord

________ 6. statement tentatively explaining observations

________ 7. plane dividing body into right and left halves

________ 8. expression comparing two numbers or quantities by division

________ 9. relationship in which X increases as Y decreases

________ 10. position of thigh relative to foot

________ 11. association of one or more types of specialized cells

________ 12. tissue composed of cells in a matrix

________ 13. comprises internal environment of body

________ 14. stable state of internal environment

________ 15. feature of every control system

II. Dichotomous. Circle the correct choice from the paired terms in parentheses.

1. (Versalius/Galen) is known for his anatomical drawings.
2. (Priestly/Pasteur) discovered oxygen.
3. In (deductive/inductive) reasoning, one proceeds from a general rule to a specific conclusion.
4. The scientific method is based on (deductive/inductive) reasoning.
5. A milligram is (larger/smaller) than a micro gram.
6. A (ratio/proportion) is an expression that compares two numbers or quantities by division.
7. The (independent/dependent) variable is usually represented on the horizontal axis.
8. In a (direct/inverse) relationship, as X values increase, Y values increase.
9. With respect to your elbow, your fingers are (distal/proximal).
10. (Negative/positive) feedback systems work to restore normal values of a variable, thereby exerting a stabilizing influence.

III. True/False. In the space to the left of the number write T for true or F for false.

_______ 1. Harvey believed that blood went from the right ventricle directly to the left ventricle.

_______ 2. In inductive reasoning, one goes from a specific observation to general conclusions.

_______ 3. The experimental and control groups differ with respect to only one variable.

_______ 4. One thousand milligrams equals one microgram.

_______ 5. Most cells have a diameter of about 10 microns.

_______ 6. The independent variable is usually represented on the ordinate.

_______ 7. Extending the ideal line into ranges not represented by experimental data is referred to as interpolation.

_______ 8. Your spinal cord is dorsally located in your body.

_______ 9. Muscle cells comprise a type of epithelial tissue.

_______ 10. The molecular basis for the homeostatic process is the maintenance of protein conformation.

IV. Sentence Completion. Write your response in the space provided.

1. _____________ provided the first evidence for capillaries.

2. The concept of homeostasis was first described by _____________.

3. In _____________ reasoning, one goes from a specific observation to a general conclusion.

4. The scientific method uses _____________ reasoning.

5. A _______________ is 1×10^{-9} grams.

6. The number of animals that survived to the total number tested is an example of a _______________.

7. In an inverse relationship, Y values decrease as X values _________________.

8. _____________ or cranial means "nearer to the head."

9. A(n) _______________ is formed when one or more types of specialized cells become closely associated in function.

10. Extracellular fluid has a _____________ concentration of sodium ions and _______ concentration of potassium ions than intracel lular fluid.

V. Multiple Choice. Circle the letter of the best response.

1. Which of the following pairs is mismatched?
   a. Aristotle – heart was seat of intellect
   b. Herophilus – brain was seat of intellect
   c. Erasistratus – blood was produced in the heart
   d. Galen – blood flowed through arteries
   e. Harvey – discovered circulation of blood

2. Key features of Harvey's theories of blood flow include all the following except:
   a. the heart is a pump
   b. blood flows from the right ventricle to the lungs
   c. aeration of blood occurs in the lungs
   d. blood moves in a circular path
   e. blood from the lungs flows directly into the left ventricle

3. Key features of the cell theory include all of the following except:
   a. all organisms are made up of cells and their products
   b. new cells may arise from preexisting ones or spontaneously
   c. all cells have the same fundamental chemical makeup and metabolic processes
   d. the activities and processes of the organism as a whole result from interdependent workings of groups of cells
   e. cells arise by cell division

4. The first step in the scientific method is usually:

4. The first step in the scientific method is usually:
   a. forming a hypothesis
   b. doing an experiment
   c. making observations
   d. retesting the hypothesis
   e. publishing the hypothesis

5. A milligram is ________ grams.
   a. 10^3
   b. 10^2
   c. 10
   d. 10^{-1}
   e. 10^{-2}

6. Scientific notation for 6 million micrograms written as grams is:
   a. 6×10^2
   b. 6×10^1
   c. 6×10^0
   d. 6×10^{-1}
   e 6×10^{-2}

7. In a direct relationship, if the value of the independent variable increases, the value of the:
   a. dependent variable increases
   b. dependent variable stays the same
   c. dependent variable decreases
   d. X value decreases
   e. Y value stays the same

8. A ______________ plane divides the body into right and left halves.
   a. sagittal
   b. frontal
   c. transverse
   d. coronal
   e. median frontal

9. The type of tissue in which cells are loosely arranged and separated by a matrix is:
   a. epithelial
   b. connective
   c. muscle
   d. nervous
   e. none of the above

10. Which is not a feature of a control system?
    a. sensors
    b. effectors
    c. reference points
    d. feedback
    e. none of the above

VI. Calculations

1. In the table write a ratio relating the number of grams designated as "x" to the units listed in the table. Use scientific notation.

   x # grams = 1 kilogram (kg) for example,

   10^3 grams = 1 kg

   a. ________ = 1 milligram (mg)

   b. ________ = 1 microgram (µg)

   c. ________ = 1 nanogram (ng)

   d. ________ = 1 picogram (pg)

2. Using ratios and proportions, show how you would determine the number of microliters in five gallons of soda. One quart equals 943 milliliters (mL).

VII. Complete the Table.

You weigh 120 lbs and are 5′4″ tall. One gram equals 0.035 ounces, and one meter equals 39.37 inches. Write your height and weight in the units indicated in the table. Use scientific notation.

Height 5'4"	Weight 120 lbs
kilometer _____	kilogram _____
meter _____	gram _____
centimeter _____	Milligram _____
millimeter _____	microgram _____
micrometer _____	nanogram _____
nanometer _____	picogram _____

VIII. Interpretation of Graphs.

Consider the following experiment to determine the effect of light intensity on the rate of photosynthesis as measured by the production of oxygen. A piece of Elodea is immersed in a 5% sodium bicarbonate solution in a test tube. A 100-watt lamp is placed at specified distances.

The rate of photosynthesis is determined by counting the number of oxygen bubbles released by the Elodea in 1 minute. Counts are made in triplicate, and the average number per minute is plotted on the graph.

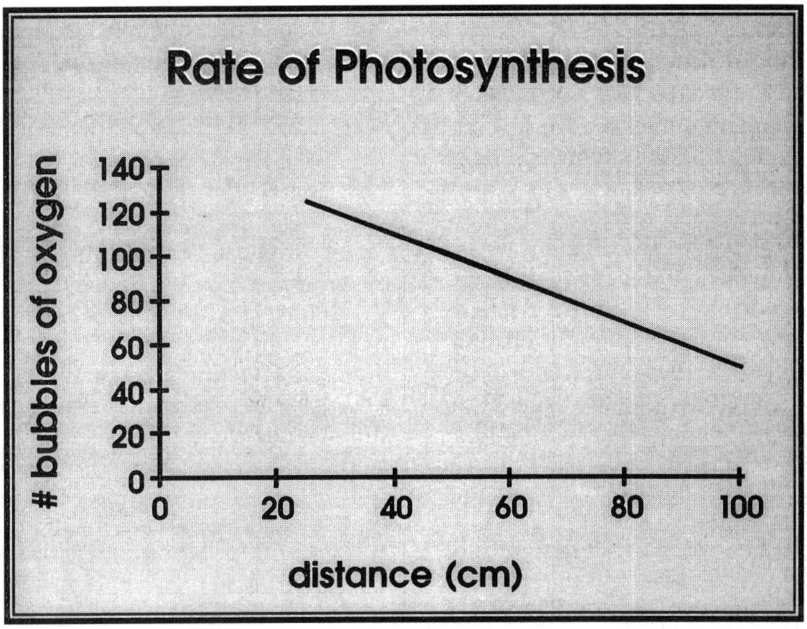

1. What is the relationship between the distance and rate of photosynthesis? ______________________

   ____

2. At 125 cm, what number of bubbles would you expect to count? ______________________

3. What is the independent variable? ______________________

4. What is the dependent variable? ______________________

5. What would be a control for this experiment? ______________________

   ____

   ____

Chapter 2

Chemical and Physical Principles

Chapter Learning Objectives

1. Explain the fundamental chemical nature of things by comparing the atoms of several simple elements and identifying common features as well as those features that distinguish elements.
2. Compare and contrast the different types of bonds, i.e., covalent, polar covalent, ionic, hydrogen, and van der Waals bonds, as to their strength, electron configuration, and formation. Provide an example of each type.
3. Consider the important features of water as a polar compound and as a solvent and describe what happens to polar covalent, ionic, and nonpolar compounds in an aqueous solution.
4. Explain the pH scale, the differences between acids and bases, and the roles of buffers.
5. Consider how the different gas laws apply to biology and write the equation expressing the relationship among amount, volume, and temperature on the pressure of a gas.
6. Describe the chemical nature of carbohydrates and how simple units can become polymerized.
7. Explain the differences between lipids and fat-like chemicals.
8. Describe how proteins are produced. (a) Write the general structural formula for an amino acid and identify the function of each component. (b) Compare the four levels of structural organization seen in many proteins regarding the shape of the polypeptide/protein, the types of bonds stabilizing each level, and the effect of pH and temperature on the levels.
9. Describe how nucleic acids are produced. (a) Distinguish a nucleotide and nucleoside. (b) Describe the structure of double helix of DNA.
10. Give the role of energy in biological processes and describe how enzymes affect activation energy for specific metabolic reactions. (a) Draw a graph expressing the relationship between substrate concentration and reaction rate and indicate Vmax and Km values. (b) Describe how pH and temperature affect enzyme activity.

Expanded Chapter Outline

I. Atoms and Elements

A. Atoms are composed of subatomic particles.

1. An **atom** is the smallest unit of an element that retains all the chemical properties of the element.
2. Atoms are composed of protons, neutrons, and electrons.
3. The number of protons, the atomic number, is unique for each element.
4. The mass number is the total number of protons and neutrons in the nucleus.
5. **Isotopes** of an element have the same atomic number but different mass numbers because they have different numbers of neutrons.

B. Ions are atoms with electrical charges.

1. An **ion** results when a neutral atom gains or loses an electron.
2. Cations and anions are positively and negatively charged ions.

C. Atoms prefer a full outermost shell.

1. Electrons are located in shells that represent different energy levels and hold different numbers of electrons.
2. The number of electrons in the outer shell determines many of the chemical properties of the atom.

II. Molecules, Groups, and Compounds

*A. A **covalent bond** is produced by electron sharing.*

1. Electrons may be shared equally (non-polar covalent bond) or unequally (polar covalent bond).

2. More than one pair can be shared.

*B. An **ionic bond** is formed by gain or loss of electrons.*

1. One atom gains one or more electrons and a second atom loses electrons to the first atom.

2. The equal and opposite electrical charges of the two atoms hold the ions together.

C. Van der Waals bonds can develop between atoms that are close together.

1. Van der Waals bonds form when a negatively charged electron cloud in one atom is attracted to the positively charged nucleus of a neighboring atom.

2. Collectively, these bonds are strong and help to stabilize three-dimensional shapes of molecules.

D. Hydrogen bonds occur between polar molecules.

1. **Hydrogen bonds** may occur within the same molecule or between molecules.

2. They result from the attraction between the partial positive charge of the hydrogen atom and the partial negative charge of an atom, such as nitrogen or oxygen.

III. Chemical Reactions: Basic Principles

A. Chemical reactions result from the breaking and forming of chemical bonds.

1. The rate of a chemical reaction depends on temperature, concentration of reactants, and the orientation of the molecules.
2. In oxidation-reduction reactions, one reactant loses electrons (oxidation), which the other reactant gains (reduction).

IV. Water and Aqueous Solutions

A. Water is a polar molecule.

1. Polar covalent bonds join the oxygen and hydrogen atoms in a water molecule.

2. Electrons are more attracted to the oxygen creating a small negative charge near the oxygen and a small positive charge near the hydrogens.

3. Hydrogen bonding between water molecules is responsible for many of water's unique properties.

B. Water is an excellent solvent.

1. Ions dissociate in aqueous solutions, and each ion is surrounded by a sphere of water molecules.

2. Uncharged polar molecules dissolve readily.

3. Non-polar compounds are poorly soluble in water and cluster together in hydrophobic "pockets" to minimize their contact with water molecules.

4. **Hydrophobic** substances (non-polar covalent) do not readily dissolve in water.

D. Concentrations of solutions are described in several ways, such as molality, molarity, or osmolarity.

E. The pH scale describes the concentration of hydrogen ions.

1. **pH** is a logarithmic expression of the concentration of H^+ ions: $\mathbf{pH = -log[H^+]}$

2. Water has a pH value of 7 and is neutral.

F. Acids and bases change the pH of solutions.

1. An **acid** is any substance whose dissociation in water releases H^+ ions. Weak acids give up H^+ ions less readily and do not dissociate completely in water.

2. A **base** is a substance that accepts H^+ ions and thus decreases the H^+ concentration in a solution.

3. Buffers are substances that accept or release

H^+ ions to maintain and minimize changes in pH in a defined range.

V. The Behavior of Gases

A. The ideal gas law relates physical properties of any gas.

1. The volume (V) of a gas is affected by temperature (T in Kelvin and Celsius), pressure (P), and amount (n = number of moles).

2. The relationship between these physical parameters is summarized in the Ideal Gas Law equation **PV = nRT,** where R is the ideal gas constant.

3. Partial pressure is directly proportional to the amount of gas present.
   a. The total pressure in a container holding a solution of gases is the sum of the partial pressures of the component gases. This is the law of partial pressures.
   b. The partial pressure of a gas is directly proportional to the moles of the gas solution it makes up.

4. Gas molecules are in continuous random motion.

VI. Macromolecules

A. Carbohydrates form branched chains.

1. Carbohydrates are classified as mono-, di-, or polysaccharides and may form compounds with proteins and lipids.

2. The general formula for a **monosaccharide** is $(CH_2O)n$, where n is the number of car bons. Monosaccharides can exist as linear chains or rings.

3. **Disaccharides** form from two monosaccha rides joined together and may exist as isomers.

4. **Polysaccharides** are formed from three or more monosaccharide units and may exist as linear or branched chains.

5. Many polysaccharides occur in covalent combination with proteins (**glycoproteins**) and lipids (**glycolipids**).

B. Lipids have low solubility in water.

1. **Lipids,** composed of carbon and hydrogen, have a lower proportion of oxygen compared with carbohydrates.

2. A **fatty acid** is a long chain of covalently linked carbon atoms to which only hydrogen atoms are attached except for a -COOH group at one end.

3. Fatty acids can be distinguished by the number of carbons and the number of carbon-carbon double bonds in the chain.

4. **Glycerol,** a three-carbon compound with three hydroxyl groups, can form bonds with the -COOH groups of one, two, or three fatty acids to form mono-, di-, and tri- acylglycerols.
   a. Glycerol may join to a phosphate group and two fatty acids to a form a **phospholipid** with **hydrophilic** and hydrophobic ends.
   b. Phospholipids are important components of cell membranes.

5. The basic structure of a steroid is four inter locking rings of carbon atoms. **Cholesterol,** an important structural component of membranes, has hydrophilic and hydrophobic regions.

6. Glycolipids occur in all membranes and are derived from **sphingosine** and monosac-charide units.

C. Proteins are unbranched chains of amino acids.

1. **Amino acids** have the same general structure.
   a. Four different chemical groups are bonded to the alpha carbon: –H, –COOH, $-NH_2$ and R.
   b. The R differs for each amino acid and determines its properties.
   c. The number and sequence of amino acids in a polypeptide chain determines its properties and functions.
   d. Amino acids are divided into three broad groups (i.e., ionic, nonionic polar, and non-polar) based on the properties of their side chains at physiological pH.

2. A protein may have four different structural levels of organization.
   a. The **primary structure** is the linear sequence of amino acids joined by peptide bonds.

b. The **secondary structure** results from spontaneous interactions among side chains and leads to a folding into a specific shape or conformation. **Alpha-helices** and **beta-sheets** are two classes of secondary structure.
c. **Tertiary structure** is formed by the folding of the alpha-helix on itself or by stable associations between amino acids located in different parts of the chain. Proteins with tertiary structure are globular.
d. Proteins with the **quaternary structure** are composed of two or more polypeptide chains held together by non-covalent interactions.

3. The secondary, tertiary, and quaternary structures are stabilized by hydrogen, ionic, and S-S covalent bonds and hydrophobic interactions between side chains.

D. Nucleic acids are long unbranched chains of nucleotides.

1. A **nucleotide** is composed of a sugar (either ribose or deoxyribose); one, two, or three phosphate groups; and a **purine** (adenine or guanine) or **pyrimidine** (cytosine, uracil, or thymine) base.

2. Nucleotides also have important roles in cell energy processes (e.g., ATP).

3. In a nucleic acid, nucleotides are joined to each other by sugar-phosphate bonds.
   a. The sugar-phosphate chain is a constant feature and is the backbone of the molecule.
   b. The sequence of bases varies, and the precise sequence represents genetic information.

4. DNA is a double helix composed of two complementary, antiparallel nucleotide chains.
   a. Base pairs are held by hydrogen bonds.
   b. In DNA, adenine pairs with thymine and cytosine with guanine.

VII. Thermodynamic Principles

A. Living organisms are highly ordered and require energy input to maintain this organized state.

1. The first principle of thermodynamics states that energy can neither be created nor destroyed.

2. The second principle states that the disorder in the universe is always increasing.

B. The direction of chemical reactions is determined by the free energy change.

1. Free energy, symbolized by G, is used to do work in the cell and is measured in kilocalories.

2. Chemical reactions involve changes in free energy.
   a. If the free energy change, delta G, of a reaction is negative, the reaction is exergonic. If delta G is positive, the reaction is energetically unfavorable and endergonic.
   b. Exergonic and endergonic reactions are coupled in the cell.

C. Enzymes decrease the activation energy for a reaction.

1. All reactions require an initial amount of activation energy to break existing bonds and create new ones.

2. Enzymes allow reactions to occur under physiological conditions.
   a. An enzyme facilitates a reaction by providing binding site(s) (active sites) on its surface for the substrate(s) to hold them in the correct orientation for the reaction to occur.
   b. Each enzyme is specific for its substrate.
   c. Enzymes are classified according to the general types of reactions they catalyze, e.g., dehydrogenases, oxidases.

3. The relationship among reaction rate, enzyme turnover, number, and substrate concentration is represented in the equation $\mathbf{V_{max} = k\,[E]}$ in which k = the turnover number, and [E] = the enzyme concentration.

4. Enzymes are affected by pH and temperature.

5. An enzyme cofactor is a non-protein component that some enzymes need to function.

6. Coenzymes are organic molecules that can serve as donors or acceptors of atoms or functional groups added to or removed from the substrate.

D. ATP acts as a free energy store.

Key Terms

acid
activation energy
alpha-helix
amino acid
anion
atomic number
base
beta-sheet
buffer
carbohydrate
catalyst
cation
cofactor
covalent bond
endergonic
enzyme
exergonic
fatty acid
glycerol
glycolipid
glycoprotein
hydrogen bond
hydrophilic
hydrophobic
interaction
ionic bond
lipid
mass number
monosaccharide
nucleic acid
nucleotide
oxidation
peptide bond
phospholipid
polypeptide
polysaccharide
protein
saturated
steroid
unsaturated

Testing Your Knowledge

I. Matching. Select the correct term from the Key Terms.

________ 1. substance that stabilizes pH

________ 2. describes energy-releasing reaction

________ 3. positively charged ion .

________ 4. lipid composed of four interlocking rings

________ 5. describes fatty acid chain with double bonds between carbons

________ 6. energy barrier in chemical reaction

________ 7. number of protons in nucleus

________ 8. secondary structure of protein

________ 9. complex of carbohydrate and protein

________ 10. weak bond between bases in DNA molecule

________ 11. proton donor

________ 12. bond between amino acids in peptide chain

II. Dichotomous. Circle the correct choice from the paired terms in parentheses.

1. The atomic (mass/number) refers to the number of protons and neutrons.
2. Negative ions are called (anions/cations).
3. Atoms with more protons than electrons are (positively/negatively) charged.
4. In oxidation an atom (loses/gains) an electron.
5. (Hydrogen/Ionic) bonds join water molecules together.
6. Fats are typically (hydrophobic/hydrophilic).
7. A high affinity of an enzyme for its substrate is predicted by a (high/low) Km.
8. The sugar in DNA is (ribose/deoxyribose).
9. Zn^{+2} may be a (coenzyme/cofactor).
10. (Endergonic/Exergonic) reactions release energy.

III. True/False. In the space to the left of the number write T for true or F for false.

_______ 1. All atoms have at least one proton.

_______ 2. Electrons with the lowest energy are closest to the nucleus.

_______ 3. Oxidation may occur without reduction.

_______ 4. At equilibrium the rates of the forward and reverse reactions are equal.

_______ 5. Saturated fats have one or more double bonds.

_______ 6. All phospholipids derive from glycerol.

_______ 7. Polar molecules are usually hydrophobic.

_______ 8. Hydrogen bonds are stronger than covalent bonds.

_______ 9. Nucleotides are found only in DNA.

_______ 10. Each active site may accommodate only one specific substrate molecule.

V. Multiple Choice. Circle the letter of the best response.

1. Which of the following is true?
   a. A proton is the smallest subatomic particle.
   b. There are always equal numbers of neutrons and electrons in an atom.
   c. The proton is positively charged.
   d. A neutron has a negative charge.
   e. Isotopes differ in atomic number.

2. The weakest type of chemical bond is:
   a. hydrogen
   b. covalent
   c. ionic
   d. Van der Waals
   e. c and d

3. A pH of 6 is ________ times more acidic than a pH of 8.
   a. 3
   b. 30
   c. 10
   d. 100
   e. 1000

4. From the ideal gas law we know that:
   a. pressure is directly proportional to volume
   b. pressure is inversely related to tempera ture
   c. volume is inversely related to pressure
   d. volume is inversely related to amount
   e. none of the above

5. Sucrose is:
   a. a monosaccharide
   b. composed of glucose monomers
   c. a protein
   d. hydrophobic
   e. an enzyme

6. Glycerol may form bonds with ______ fatty acids.
   a. 1
   b. 2
   c. 3
   d. a, b, and c
   e. a and b

7. The sequence of amino acids joined by peptide bonds represents the ______________ structure.
   a. primary
   b. secondary
   c. tertiary
   d. quaternary

8. Which is not true for cofactors?
   a. They may be ions like zinc and copper.
   b. They may be organic molecules.
   c. They are unaffected by the reaction.
   d. They are necessary for enzyme activity.
   e. c and d

9. In DNA, A pairs with:
   a. G
   b. C
   c. T
   d. U
   e. none of the above

10. Maltase is a(n):
    a. carbohydrate
    b. protein
    c. amino acid
    d. lipid
    e. nucleotide

V. Complete the Table.

Element	Symbol	Atomic #	# Protons	# Neutrons	Isotopes
hydrogen					
	^{12}C				
nitrogen					
	O^{2-}				
calcium					

VI. Visual #1.

a) How many bonds can this atom form?
____

b) How many protons are present? _______

c) What is the atomic number? ________

d) How many neutrons are present? _______

e) What type of bond(s) can this atom form?
____

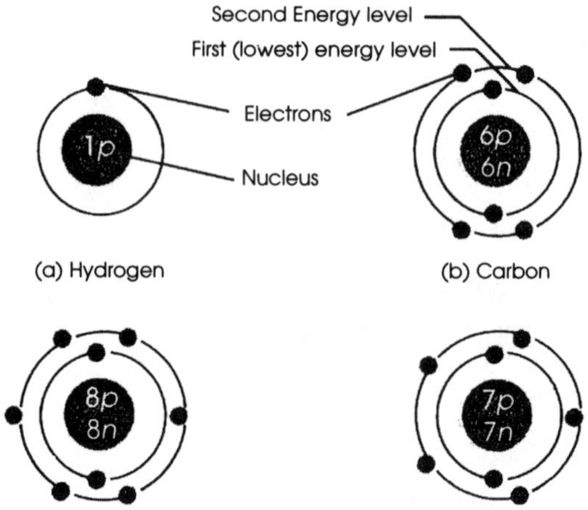

(a) Hydrogen (b) Carbon (c) Oxygen (d) Nitrogen

Chapter 3

The Structure and Function of Cells

Chapter Learning Objectives

1. Restate the cell theory and explain why it is one of the most basic biological theories.
2. Diagram a cell and label all of its major parts.
3. List each cellular organelle and its function. (a) Diagram information flow in the cell. (b) List the organelles and their involvement in the production, processing, and packaging of a secretory protein.
4. Present the cellular environment and describe the role of the plasma membrane.
5. Support the statement: The nucleus is the control center for the cell.
6. Summarize the cell cycle and the events of mitosis and account for abnormal cell cycles.
7. Illustrate how cell culture could be used to solve some of the mysteries surrounding cancer or normal cell metabolism.
8. Indicate how different kinds of microscopes can be used to study cells.
9. Point out how radioisotopes can be used to study cells.
10. Detail some of the new approaches to the study of the cell.

Expanded Chapter Outline

I. Overview: Cell Processes and Products

A. *Instructions for controlling cell functions are in the nucleus.*

1. The genes in the DNA encode a set of instructions that the cell uses to make macromolecules needed to carry out its functions in the body and to reproduce itself.
   a. DNA is made, stored, and decoded in the **nucleus**.
   b. DNA encodes instructions for making proteins, tRNA, and rRNA.

2. Information flows from genes to RNA molecules, which, along with ribosomes in the cytoplasm, play a major role in the production of proteins.
   a. Proteins made on free **ribosomes** participate in internal cell processes.
   b. Proteins made on ribosomes attached to the ER are either exported from the cell or become part of organelles or membranes.

3. DNA replicates itself to pass information to daughter cells.

B. *Energy obtained from food is used to support life processes.*

1. **Organelles** function to convert energy into a useful form to support processes such as growth and reproduction.

2. Mitochondria are the major sites for ATP synthesis.

C. *Membranes protect the cell interior and divide it into different compartments.*

1. The cell's membrane system separates the cytoplasm from the extracellular environment and maintains the intracellular environment.

2. The plasma membrane, a bilayer of phospholipids with embedded proteins, provides a barrier to the free movement of materials into and out of the cell.

3. Intracellular compartments in the organelles permit separation of synthetic and degradative reactions and allow for intra cellular processing, sorting, and packaging of synthesized molecules.
   a. Each organelle has a defined structure and function that are intimately related.
   b. An organelle's specialized function is carried out by enzymes located within the organelle or in its membrane(s).

II. Cell Compartments and Their Functions

A. The nucleus is the library of the cell.

1. **Chromatin**, a complex of DNA and histone, is organized into nucleosomes, which exhibit higher levels of packaging, such as chromatin fibers, looped domains, and chromosomes.

2. Each cell has 23 pairs of chromosomes.

3. Three functional types of **RNA** (mRNA, tRNA, and rRNA) play a role in the protein synthesis.

4. Exchange of materials between the nucleus and cytoplasm occurs through the nuclear **pores.**

B. The cytosol and organelles are the sites of degradation and energy release.

1. The cytosol contains precursor molecules for macromolecules, enzymes, and other granules and droplets.

2. The amount of each of the two functional regions of the **endoplasmic reticulum** (ER), smooth and rough, vary with the function of the cell.
   a. Smooth ER is involved in the synthesis and secretion of lipids and lipoproteins that may also become components of cell membranes or released from the cell.
   b. Rough ER has attached ribosomes and is involved in synthesis of proteins to be secreted from or incorporated into the membranes and organelles of the cell.
   c. Proteins synthesized on the rER are modified by the addition of carbohydrates in the rER and then passed via a transition vesicle to the Golgi complex.

3. The **Golgi complex** processes and packages macromolecules into vesicles ready for transport to the cell surface and the cytosol.

4. Membrane recycling involves the processes of exocytosis and endocytosis and occurs among the ER, Golgi, lysosomes, and plasma membrane.

5. **Lysosomes** and peroxisomes digest food and waste.
   a. Primary lysosomes contain digestive enzymes and arise from the Golgi complex. Secondary lysosomes form after the fusion of primary lysosomes with endocytic vesicles.
   b. Lysosomal enzymes digest material in endocytic vesicles and cellular organelles and matter when necessary in a process called autophagy.
   c. **Peroxisomes** contain enzymes involved in detoxification reactions.

6. The **mitochondria** are the primary sites for aerobic respiration and ATP synthesis.
   a. Their number and position in the cell vary according to the cell's need for ATP and its function in the body.
   b. The mitochondrion has a double membrane, DNA, and ribosomes.
   c. The inner membrane and matrix contain enzymes and other molecules used to oxidize fuel molecules and to make ATP.

7. The **cytoskeleton** is responsible for maintaining cellular shape and the positions of organelles and for movement of materials and organelles within the cell.
   a. It is composed of different types of protein fibers (**microtubules, microfilaments,** and intermediate filaments), which differ in their composition, structure, function, and location.
   b. Microtubules play a major role in organizing the other filaments in the cytoskeleton.
   c. Myosin and actin filaments form part of the contractile machinery in muscle and nonmuscle cells.
   d. Microfilaments and microtubules are dynamic.

III. The Life Cycle of the Cell

~ The different types of cells in the body vary as to their patterns and rates of division.

~ The cell cycle, the time period from the start of one division to the start of the next division, consists of two main phases: interphase and cell division.

A. Cell growth and DNA replication occur during interphase.

1. **Interphase** is divided into three phases: G_1, S (synthetic stage), and G_2.

2. During interphase, the cell carries out normal metabolic activities and duplicates DNA.

3. G_1 is the most variable phase timewise.

B. *Cell division occurs during mitosis and cytokinesis.*

1. **Mitosis** occurs in four stages: prophase, metaphase, anaphase, and telophase, which are distinguished on the basis of chromosomal appearance and activity.

2. **Cytokinesis**, the division of the cytoplasm, occurs during telophase.

C. *The cell cycle "engine" controls division.*

1. Maturation promoting factor (MPF) induces a cell to enter M from G_2.

2. The cell cycle is regulated by the periodic activation and inactivation of MPF.
   a. **Cyclin B** is a component of the "engine."
   b. Other cyclins are important in the control of DNA replication in S and in moving the cell from G_1 to S.

D. *Cancer cells have abnormal cell cycles and do not respond to normal controls of cell division and growth.*

IV. Research Techniques in Cell Physiology

A. *Human cells grow in the laboratory.*

1. Most cells from mammalian tissues can grow and divide in a culture medium containing the required nutrients.

2. Normal cells in culture have a finite lifetime and exhibit contact inhibition in contrast to cancerous cells.

B. *Microscopes allow visualization of cells and organelles.*

1. With the **light microscope**, the specimen is magnified about 1000 to 1500 times.

2. With the electron microscope, the specimen is magnified more than 100,000 times.

3. **Cytochemistry** is used in combination with microscopy to produce a color or enhanced contrast at the sites of specific molecules.

C. *Autoradiography detects the location of specific molecules.*

D. *Antibodies are used to detect, isolate, or measure specific molecules.*

1. Labeled antibodies specific for an antigen are used to locate the antigen within the cell or tissue.

2. **Immunocytochemistry** uses a fluorescent antibody to reveal the location of the targeted molecule.

3. Radiolabeled antibodies are also used to locate targeted molecules.

4. **Radioimmunoassays** are used to quantify the amount of targeted molecule in the cell.

5. **Affinity chromatography** takes advantage of the specificity of antigens and antibodies to isolate antibodies from body fluids passing through a matrix with bound antigen.

6. **Monoclonal antibodies** are produced by hybridomas formed by the fusion of antibody-producing and cancer cells.

E. *HPLC is important for rapid amino acid sequencing.*

F. *Cell organelles are isolated by cell fractionation.*

1. **Cell fractionation** allows for the separation of the different organelles so that each can be studied in a controlled "cell-free" system.

2. Organelles are separated by differential or density gradient **centrifugation.**

G. *New molecules are formed by genetic engineering.*

1. DNA from different organisms can be combined to form a recombinant molecule.

2. A gene can be isolated, cloned, and amplified to study its function and structure.

Key Terms

affinity chromatography
antibody
antigen
autophagy
autoradiography
cell fractionation
cell culture
chromatin
chromosome
cytochemistry
cytokinesis
cytoskeleton
endocytosis
exocytosis
gene
Golgi complex
homogenization
interphase (G_1, S, G_2)
light microscope
lysosome
microfilament
microtubule
mitochondria
mitosis
monoclonal
myosin filament
neoplasm
nuclear envelope
nucleolus
nucleus
plasma membrane
radioimmunoassay
radioisotopes
ribosome
rough ER (rER)
smooth ER (sER)

Testing Your Knowledge

I. Matching. Select the correct term from the Key Terms.

________ 1. complex of DNA and histones

________ 2. composed of double membrane with pores

________ 3. major site of ATP production

________ 4. longest phase of cell cycle

________ 5. site of synthesis of proteins used within cell

________ 6. site of synthesis of proteins exported from cell

________ 7. site of synthesis of ribosomal RNA

________ 8. cell without a nucleus

________ 9. contains digestive enzymes, involved in autophagy

________ 10. process by which membrane vesicles leave cell

________ 11. process by which membrane-enclosed materials enter a cell

________ 12. site of lipid synthesis

________ 13. site of processing and packaging proteins for export

________ 14. abnormal growth

________ 15. composed of different types of protein fibers and functions in maintaining cell shape and internal movement

II. Dichotomous. Circle the correct choice from the paired terms in parentheses.

1. DNA replication occurs during (interphase/prophase).
2. Hydrogen peroxide would be destroyed by enzymes located in the (peroxisomes/lysosomes).
3. The process by which secretory proteins leave the cell is (endocytosis/exocytosis).
4. Proteins made on (free ribosomes/rough ER) would become part of cellular membranes.
5. (Very few/Many) mitochondria would be found in muscle cells.
6. In cytokinesis the (nucleus/cytoplasm) divides.
7. Cloned cells are by definition (identical/not identical).
8. An (antibody/antigen) stimulates the immune system.
9. ATP production occurs mainly in the (mitochondria/cytoplasm).
10. Each chromosome contains (one/many) molecule(s) of DNA.

III. True/False. In the space to the left of the number write T for true or F for false.

________ 1. RNA synthesis occurs in the cytoplasm.

________ 2. Materials move through the nuclear membrane by endocytosis and exocytosis.

________ 3. Each organelle has a unique set of enzymes.

________ 4. Smooth ER is the site of protein synthesis.

_______ 5. Lysosomes contain synthetic enzymes.

_______ 6. Microtubules are found in the cell's cytoskeleton.

_______ 7. Nuclear division occurs during anaphase.

_______ 8. Cytokinesis occurs during interphase.

_______ 9. Ultraviolet light is used to reveal fluorescent dyes attached to mole cules in a cell.

_______ 10. Mitochondria may be isolated using cell fractionation.

IV. Sentence Completion. Write your response in the space provided.

1. A _____________ cell has a nucleus.

2. The sequence of nucleotides encoding information for a protein is called ___________.

3. Proteins for export may be modified in the __________ and ____________.

4. Proteins for export are packaged in the ____________.

5. Large molecules enter the cell by the process of _______________.

6. Secretory proteins leave the cell by the process of _____________.

7. The sum total of all genes in a cell is called the ____________.

8. The smooth ER synthesizes ____________.

9. The structure of microscopic finger-like projections from the surface is due to _______________ located therein.

10. Radioactive isotopes are used in ______________ and _____________.

V. Multiple Choice. Circle the letter of the best response.

1. The nucleus of the cell:
   a. contains just DNA
   b. contains DNA and RNA
   c. is separated from the cytoplasm by one membrane
   d. manufactures proteins
   e. is the site of cellular metabolism

2. The most common protein in the nucleus is:
   a. chromatin
   b. histone
   c. tubulin
   d. myosin
   e. none of the above

3. The type of RNA that carries the information for a protein is:
   a. mRNA
   b. tRNA
   c. rRNA
   d. rDNA
   e. both a and b

4. Microtubules are located in the:
   a. Golgi complex
   b. peroxisomes
   c. cytoskeleton
   d. ribosomes
   e. mitochondria

5. Microfilaments would be expected in cells of the:
   a. kidney
   b. liver
   c. intestine
   d. skin
   e. all of the above

6. DNA synthesis occurs in:
   a. interphase
   b. metaphase
   c. anaphase
   d. telophase
   e. prophase

7. Chromosomes separate during:
   a. interphase
   b. metaphase
   c. anaphase
   d. telophase
   e. prophase

8. The site of lipids synthesis is the:
   a. mitochondria
   b. rER
   c. sER
   d. Golgi
   e. lysosome

9. To look at the structure of a mitochondrion one would use:
   a. light microscope
   b. electron microscope
   c. radioimmunoassay
   d. monoclonal antibodies
   e. all of the above

10. A secretory vesicle leaves the cell by:
    a. endocytosis
    b. exocytosis
    c. diffusion
    d. mitosis
    e. none of the above

VI. Complete the Table.

Cellular Structure	Molecular Composition	Function
nucleus		
nucleolus		
rER		
sER		
ribosome		
Golgi complex		
lysosome		
peroxisome		
mitochondria		
cytoskeleton		
microfilaments		
microtubules		

VII. Problem

A piece of skin is placed in a medium containing a radioactively labeled amino acid and left there for hours. The tissue is homogenized, the cells fractionated, and the organelles separated using differential centrifugation. Please answer the questions on the following page.

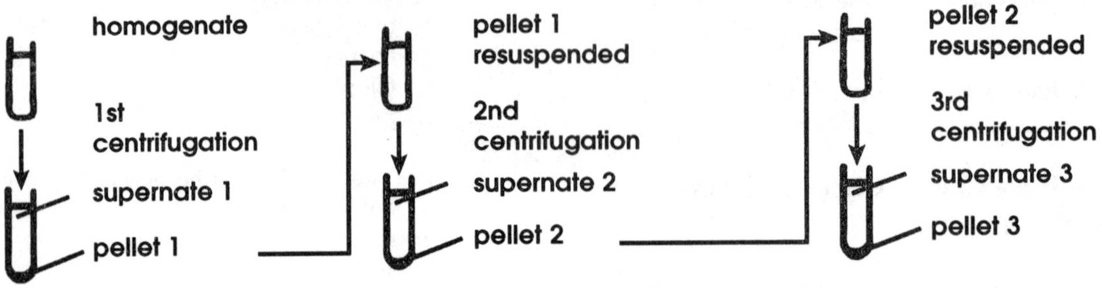

a. Which technique(s) would you use to determine the contents of pellets 1, 2, and 3?

b. In which pellet would you find digestive enzymes?

c. Which technique(s) would you use to determine the location of the radioactively labeled amino acid?

d. Where do would you expect to detect radioactivity?

e. In which pellet would you find DNA?

f. Where would you find RNA?

VIII. Label the Figure.

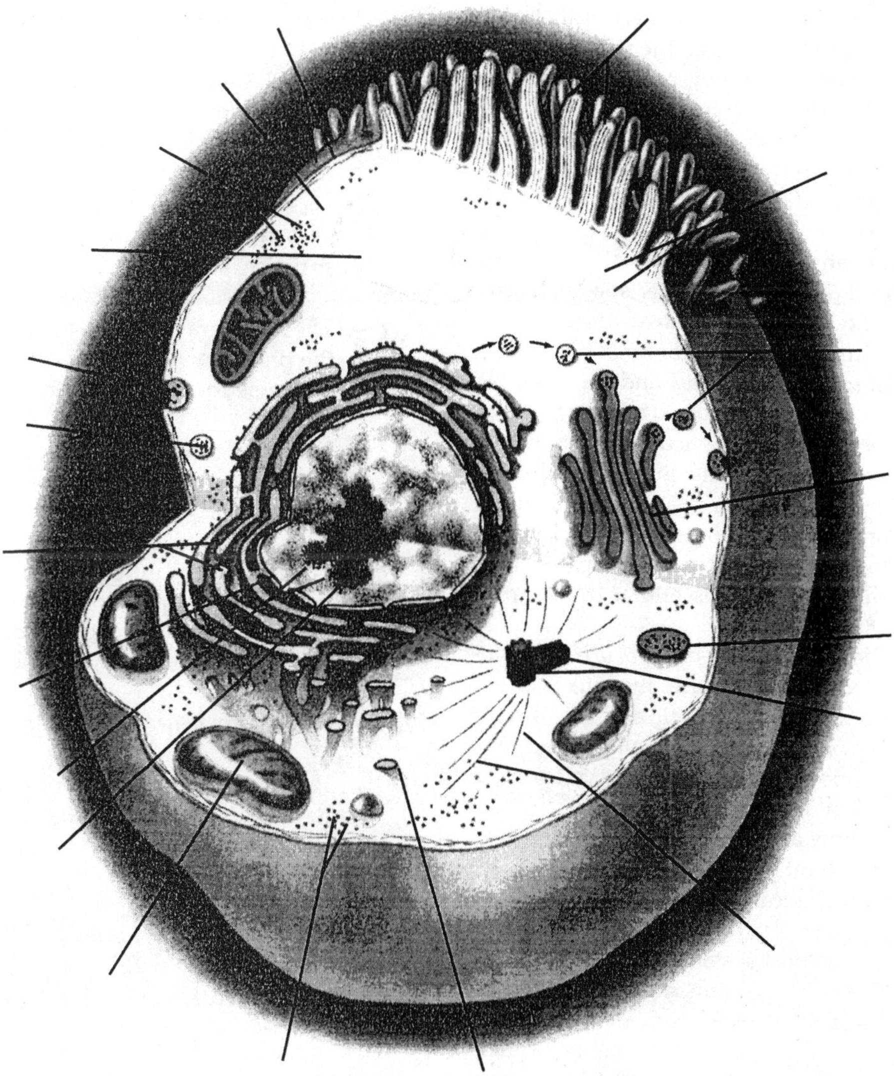

Chapter 4

Transport Through the Cell Membrane

Chapter Learning Objectives

1. Describe the molecular structure of the plasma membrane.
2. Explain the fluid mosaic concept.
3. Explain how membrane proteins function in transport of ions through the membrane.
4. Give the roles of proteins and carbohydrates in the plasma membrane.
5. Detail the process of simple diffusion focusing on the characteristics of the solute. Consider how gradients affect the movement of molecules.
6. Compare and contrast simple diffusion and facilitated diffusion.Consider specificity for solute and the role of intrinsic proteins, effect of molecular size, charge, lipid solubility of solute on the rate, and the rates of these processes and sources of energy.
7. Describe the process of active transport using the epithelial cells lining the proximal renal tubule. Consider the functional and structural specialization seen in membranes of epithelial cells and how each specialization contributes to transcellular transport of water and solutes.
8. Relate how osmolarity of surrounding solutions affects cells. Consider the effects of hypoosmotic, hyperosmotic and iso-osmotic solutions on the volume of cells placed in these solutions and describe how cell volume is maintained.
9. Indicate how and why membrane fragility is an important physiological consideration.
10. Give the biological significance of the Na^+/K^+/ATPase pump. (a) Compare active and passive transport systems with respect to cellular energy requirements, role of intrinsic proteins, direction of movement of solute, and effect on cellular volume. (b) Compare primary and secondary active transport processes.

Expanded Chapter Outline

I. Overview of Plasma Membrane Functions

~ The plasma membrane plays an important role in cell-cell communication.

~ The plasma membrane functions as a selectively permeable barrier, thereby allowing the composition of the cytosol to be different from its surrounding environment.

II. The Components and Structure of the Membrane

~ The structural backbone of a membrane is the phospholipid bilayer, in which proteins are embedded or loosely associated.

~ The functional properties of the membrane are due to the proteins, whose mass varies from 18%–76%.

~ The relative quantities of lipids and proteins vary for specific membranes and usually reflect some functional aspect of the membrane around the cell or organelle.

A. *Membrane lipids are amphipathic molecules.*

1. Phospholipids, cholesterol, and glycolipids have hydrophobic and hydrophilic regions and form a bilayer in an aqueous environment.

2. Phospholipids can rotate about their axes, undergo flexing motions, and exchange places rapidly within the same layer by lateral diffusion but "flip-flop" very slowly. Cholesterol stabilizes hydrocarbon chains of the phospholipids.

3. Different types of phospholipids are not distributed equally between the halves of the bilayer.

4. The membrane's fluidity is like olive oil and is important to normal functioning.

B. *Most membrane proteins penetrate the lipid bilayer.*

1. Hydrophobic regions of intrinsic proteins are clustered in the hydrophobic interior of lipid bilayer, and hydrophilic regions are located at aqueous interfaces.

2. Many intrinsic proteins can undergo rapid lateral diffusion and axial rotation but probably do not "flip-flop."

3. Extrinsic (peripheral) proteins are associated with the membrane primarily by hydrogen bonding and electrostatic interactions.

4. The asymmetrical distribution and types of membrane proteins are related to cell and membrane function.

C. *Carbohydrates are attached to lipids and proteins on the surface of the membrane.*

1. The hydrophilic regions of glycoproteins and glycolipids interface with the aqueous environment.

2. Carbohydrates are minor membrane components but play a role in cell-cell recognition processes and defining receptor sites.

3. They are unequally distributed between the two layers, and glycolipids are found only in the eternal half.

III. Gradients and Movement

A. *A pressure gradient drives blood flow.*

1. Gradients exist in many forms (e.g., pressure, electrical, chemical) and play important roles in the movement of substances into cells and throughout the body.
   a. The greater the difference between the two points, the greater the flow.
   b. Flow is inversely related to resistance.

2. The combined effects of pressure and resistance on flow (F) are represented by the equation: **F = (P_1 minus P_2)/R**. P_1 and P_2 represent the pressures at the starting and final points. R is resistance.

B. *A concentration gradient drives diffusion.*

1. In diffusion of solutes in liquid solvents, particles of solute move by Brownian motion down a concentration gradient until a dynamic equilibrium is reached.
   a. The steeper the gradient, the greater the net movement.
   b. **Flux** describes the rate of movement of solute and is expressed as amount of solute moving per unit time, e.g., mmol/min.
   c. Flux is expressed by **Fick's Law: J = minus DA times dc/dx**. J = flux; D = diffusion coefficient, which varies with specific solute and solvent; A = area through which diffusion occurs; dc/dx = driving force or concentration gradient; minus = directional movement from high to low.

2. The **diffusion** of gas molecules is the same as that for solid or liquid solute particles.

3. In a mixture of gases, each diffuses down its concentration gradient.

IV. Movement of Substances Through the Membrane

~ Substances pass through selectively permeable cell membranes by active or passive transport processes.

~ The lipid bilayer prevents the diffusion of polar or charged molecules except water.

A. ***Passive transport** is spontaneous movement down a gradient.*

1. Passive transport processes do not require cellular energy and occur in response to concentration gradients.

2. Simple diffusion of lipid-soluble substances depends on molecular motion.
   a. The rate of diffusion of a solute through a membrane depends on the thickness of the membrane, the lipid solubility of the solute, its molecular size and charge, and **diffusion coefficient, D.**
   b. The **permeability coefficient, P,** varies with the specific solute and type of cell membrane.
   c. Fick's law can be adapted to represent the rate of solute diffusion across a membrane: **J = minus PA [C_1 minus C_2].**

C_1 and C_2 = solute concentrations on either side of the membrane; A = area through which diffusion occurs; **J = flux** .

d. The electrical gradient across the membrane affects the diffusion of ions such as sodium and potassium.

e. The relationships between rate of diffusion and molecular size, charge and lipid solubility are summarized in Overton's Rules: (1) The permeability of cell membranes to small non-polar solutes is directly proportional to lipid solubility. (2) The permeability of cell membranes to polar solute is inversely proportional to molecular size.

f. Diffusion trapping involves the formation of a charged molecule after the neutral form of a weak base has passed through a membrane.The charged form is not able to pass back through the membrane.

3. **Osmosis** is simple diffusion of water through a membrane.

a. Water moves from a dilute solution or solution with low osmotic pressure to a concentrated or high **osmotic pressure.**

b. Water flow through the membrane is affected by the osmolarity of the cytosol and the cell's external medium.

c. For physiological solutions, osmotic concentration is expressed in units called **milliosmoles,** which are equal to n times millimoles; n equals the number of particles derived from each molecule of solute.The relationship applies only for dilute solutions.

d. Osmotic pressure is related to solute concentration by the **van't Hoff law,** which is described by the equation π = **RTC.** π = osmotic pressure, R = ideal gas constant, T = temperature (K), and C = total solute concentration expressed as osmoles of solute per liter of solution or osmolarity.

e. **Osmotic fragility** is a term describing an erythrocyte's response to osmotic stress and represents the osmolarity of NaCl causing 50% of erythrocytes to lyse.

f. Osmotic fragility can be used as a test for diseases affecting erythrocyte behavior.

4. Lipid-insoluble substances may enter the cell by **facilitated diffusion.**

a. The passage of some large, water-soluble molecules, such as D-glucose and amino acids, involves intrinsic membrane proteins acting as carriers or channels.

b. Facilitated diffusion is a rate-limited, highly specific process and can be blocked by competitive inhibitors.

c. It involves a reversible binding of the solute with carrier proteins in the membrane.

d. **Hydrophilic channels** composed of intrinsic proteins provide a way for charged ions, such as K^+ and Na^+, to cross the membrane. These channels can be regulated by "gates."

e. Facilitated diffusion is more rapid than simple diffusion but is rate-limited by the availability of binding sites on the carrier proteins for the solute.

f. The process of facilitated diffusion and the effect of competitive inhibitors can be described by kinetic parameters such as V_{max} and K_m.

B. Active transport is energized movement against a gradient.

1. Active transport processes require cellular energy and carrier proteins and result in the movement of solute against the prevailing gradient.

2. In primary active transport, the movement of the solute is directly coupled to an energy-yielding reaction.

a. The **sodium-potassium ATPase pump** is an example of primary active transport.

b. The breakdown of ATP is tightly coupled to the transport of Na^+ out of and K^+ into the cell.

3. Secondary active transport refers to the movement of the solute not coupled directly to energy-yielding reactions.

a. A transport is driven by energy stored in the electrochemical gradient of another solute, e.g., Na^+.

b. The mechanism of Na^+−coupled transport systems is believed to involve a two-site carrier that binds both Na^+ and the solute.

C. Transport processes and cell volume are interdependent.

1. Intracellular osmotic pressure and volume are regulated by the Na+/K+/ATPase pump.

a. The rates of diffusion of Na^+ and K^+ across the membrane are matched by the pumping rates of the Na^+/K^+/ATPase pump.

b. The plasma concentration of Na^+ is regulated closely to prevent large shifts in plasma osmolarity that would affect intracellular volume.

2. Many cells correct volume changes by activating pathways that increase the efflux of K^+ and Cl^- .

3. The decrease in intracellular osmolarity stimulates an efflux of water.

4. Some cells respond to a cell volume decrease by increasing the influx of solutes such as Na+ and Cl2.

V. Epithelial Transport

A. *Epithelial cells are organized asymmetrically.*

1. Transport systems are asymmetrically distributed between the apical and basolateral plasma membranes.
   a. The organization of transport systems allows for transcellular transport of water and solutes.
   b. **Na^+—coupled symport systems** for the entry of glucose, amino acids, and phosphate are located in the apical membrane. Na^+ exits the cell through the action of the Na^+/K^+/ATPase pump located in the basolateral membrane.
   c. Transcellular transport achieves net movement from the lumen at one side of the cell to blood vessels on the other side. It can be regulated by hormones.
   d. Water movement across epithelia is always passive and driven by an osmotic gradient.

2. **Junctional complexes** prevent lateral diffusion of membrane proteins in epithelial cells and mixing of proteins in the apical and basolateral surfaces.
   a. Intercellular complexes such as **tight junctions** prevent leakage between epithelial cells.
   b. "Leaky" junctions provide a **paracellular pathway** for water and solutes across the epithelium. Movement of ions via this pathway is passive and along electro chemical gradients.

3. The surface areas of the apical and basolateral membranes are increased by microvilli and extensive infoldings, respectively.

Key Terms

active transport
amphipathic
antiport
asymmetry
back flux
carrier protein
extrinsic protein
channel protein
cholesterol
competitive inhibition
diffusion
diffusion trapping
electrochemical gradient
$F = (P_1 - P_2)/R$
facilitated diffusion
flip-flop
flow
flux
glycolipid
glycoprotein
gradient
hyperosmotic
hypo-osmotic
intrinsic protein
iso-osmotic
J = 2DA 3 dc/dx
$J = 2PA[C_{12}\ C_2]$
junctional complex
lipid bilayer
micelle
Na^+/K^+/ATPase pump
osmosis
osmotic pressure
Overton's rules
osmolarity
osmole
paracellular transport
passive transport
permeability coefficient
phospholipid
specificity
symport
transcellular transport
$\pi = RTC$

Testing Your Knowledge

I. Matching. Select the correct term from the Key Terms.

_________ 1. movement of molecules down concentration gradient

_________ 2. movement of molecules against concentration gradient

_________ 3. having hydrophilic and hydrophobic regions

_________ 4. involves carrier proteins and movement down concentration gradient

_________ 5. refers to unequal distribution of proteins in membrane

_________ 6. describes cotransport of Na^+ and glucose

_________ 7. describes protein loosely associated with membrane

_________ 8. protein complexed with carbohydrate

_________ 9. movement of water through mem brane

_________ 10. term describing carrier and channel proteins

_________ 11. indirectly regulates water flow into and out of cell

_________ 12. structural backbone of membrane

_________ 13. membrane component affecting fluidity

_________ 14. lipid complexed with carbohy drate charge

_________ 15. describes relationship between osmotic pressure and osmolarity

II. Dichotomous. Circle the correct choice from the paired terms in parentheses.

1. (Intrinsic/Extrinsic) proteins penetrate the bilayer.

2. (Carbohydrates/Proteins) are a major component of membranes.

3. The larger the difference in pressure between two points, the (greater/lesser) the flow.

4. If resistance increases, flow (decreases/increases).

5. (Passive/Active) transport requires expenditure of cellular energy.

6. The rate of solute diffusion across a membrane increases/decreases) as the area increases.

7. A 30 mmol/liter solution of glucose has a (higher/lower) osmotic pressure than a 30 mmol/liter NaCl solution.

8. Osmotic pressure (increases/decreases) as solute concentration increases.

9. The sodium-potassium pump is found in the (apical/basolateral) membrane of a cell lining the intestine.

10. The sodium-potassium pump transports sodium (into/out of) the cell.

III. True/False. In the space to the left of the number write T for true or F for false.

_______ 1. Amphipathic molecules have both hydrophilic and hydrophobic regions.

_______ 2. Cholesterol destabilizes the lipid bilayer.

_______ 3. Phospholipids and glycoproteins move readily from the inner to the outer layer of the lipid bilayer.

_______ 4. Phospholipids and proteins are symmetrically distributed in the lipid bilayer.

_______ 5. Extrinsic proteins are usually not amphipathic.

_______ 6. Diffusion occurs only through a selectively permeable membrane.

_______ 7. Water is actively transported across membranes.

_______ 8. The osmotic pressure of pure water is zero.

_______ 9. The kinetics of facilitated diffusion resembles those for an enzyme.

_______ 10. After a period of time a cell placed in a solution of 300 mOsm NaCl and 60 mOsm urea eventually bursts.

IV. Sentence Completion. Write your response in the space provided.

1. Junctional complexes prevent ____________ transport across an epithelium.

2. The surface area of the apical surface of an epithelial cell is increased by ____________.

3. The lipids and proteins of the plasma membrane are capable of ______________ diffusion, but not ____________.

4. The fluidity of the plasma membrane is like ______________.

5. A membrane actively transporting molecules would have a large number of ______________ near the membrane.

6. _______________ play a role in cell-cell communication and recognition.

7. _____________ describes the rate of movement of solute molecules.

8. Facilitated diffusion is a form of _________ transport.

9. Osmotic pressure is directly proportional to _____________ if temperature is constant.

10. Movement of solute against a concentration gradient is ___________ transport.

V. Multiple Choice. Circle the letter of the best response.

1. Which characteristics apply to both membrane lipid and proteins?
   a. asymmetrical distribution
   b. amphipathic
   c. can form complex with carbohydrate
   d. can undergo lateral diffusion
   e. all of the above

2. Which characteristics apply to both simple and facilitated diffusion?
   a. specificity
   b. rate-limited
   c. involves intrinsic proteins
   d. have Vmax and Km values
   e. none of the above

3. Which of the following molecules would move most rapidly through the cell mem brane?
   a. Na^+
   b. K^+
   c. glucose
   d. oleic acid
   e. glycerol

4. Erythrocytes are placed in a 400 mOsm saline solution. Which of the following statements is true?
   a. The erythrocytes are iso-osmotic to the solution.
   b. The saline solution is hypo-osmotic to the erythrocytes.
   c. Water will enter the erythrocytes.
   d. The cells will actively transport water out of the cytosol.
   e. The erythrocytes will lose volume but not surface area.

5. Which of the following statements is true?
   a. Diffusion requires the breakdown of ATP.
   b. Facilitated diffusion uses carrier proteins.
   c. Simple diffusion is rate-limited by the concentration of carrier proteins.
   d. Simple diffusion is a form of active transport.
   e. none of the above

6. Which molecule would most likely pass by simple diffusion through the plasma membrane?
   a. small lipid-soluble molecule
   b. small anion
   c. small cation
   d. small polar molecule
   e. large non-polar molecule

7. The sodium-potassium pump:
   a. transports sodium into and potassium out of the cell
   b. transports sodium out of and potassium into the cell
   c. transports potassium and sodium into the cell
   d. transports potassium and sodium out of the cell
   e. none of the above

8. The flow of blood in arteries and veins is:
   a. directly proportional to resistance
   b. inversely proportional to pressure
   c. inversely related to resistance
   d. directly proportional to pressure
   e. c and d are correct

9. The rate of solute diffusion across a membrane is:
   a. inversely related to the permeability coefficient and area
   b. inversely related to the area of diffusion only
   c. directly proportional to area and permeability coefficient
   d. constant over time
   e. a and d are correct

10. In the presence of an inhibitor to the sodium-potassium pump, which of the following would occur?
    a. glucose transport would increase
    b. potassium ions would accumulate in the cell
    c. sodium ions would accumulate in the cell
    d. the electric potential across the cell membrane would become more negative
    e. all of the above

VI. Problem

This beaker contains 400 mOsm glucose solution, while the bag inside of it contains 300 mOsm glucose solution

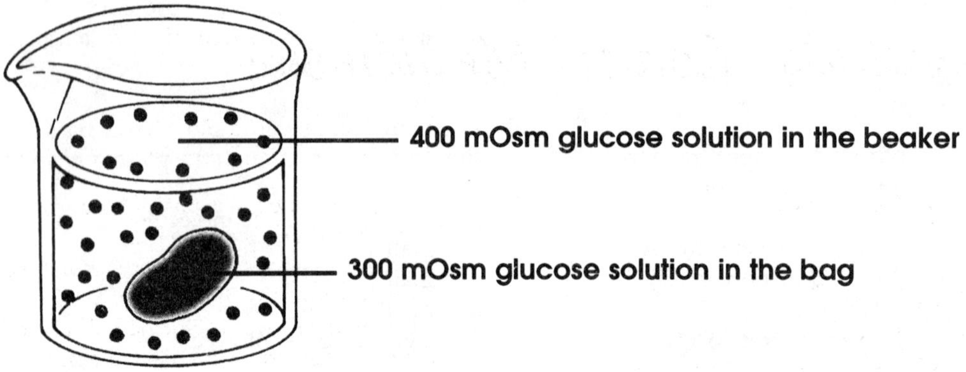

1. The bag is constructed from a semipermeable membrane that is permeable to water but not glucose molecules.
   a. In which direction is there initially a net movement of water?
   b. Using the terms hyperosmotic, hypo-osmotic, and iso-osmotic, what is the solution in the beaker relative to the solution in the bag?
   c. What is the solution in the bag relative to the solution in the beaker?

2. Suppose a 50-mOsm NaCl solution is added to the bag.
   a. In which direction will there be a net movement to water?
   b. What is the solution in the beaker relative to that in the bag?
   c. What is the solution in the bag relative to that in the beaker?

Chapter 5

Cellular Control Mechanisms

Chapter Learning Objectives

1. Give a detailed description of the chemical nature of a DNA molecule.
2. Compare and contrast DNA and RNA. Consider the types of sugars, bases, shapes, post-synthetic modifications, and functions.
3. Explain how the DNA molecule functions. (a) Consider the processes of semiconservative DNA replication and transcription and the role of signals and polymerases in these processes. (b) Consider the genetic code and DNA repair mechanisms.
4. Describe the process of protein synthesis. (a) Consider the roles of the different RNAs and enzymes. (b) Describe the events of the three stages of translation: initiation, elongation, termination.
5. Distinguish among template DNA, messenger RNA (mRNA), transfer RNA (tRNA), and ribosomal RNA (rRNA). Describe their roles in protein synthesis.
6. Indicate how a ribosome functions.
7. Outline how some of the different controls over gene expression work. Consider the levels at which genes may be controlled.
8. Explain how hormones function at the cellular level. (a) Consider local chemical mediations and hormones and the role of receptors. (b) Consider the duration of response in the target cells and the hormone's mechanism of action in eliciting a response from the target cell. (c) Consider the roles of first and second messengers in generating a physiological response in the target cell.
9. Summarize the types of interactions between adjacent cells and distant cells.

Expanded Chapter Outline

I. An Intracellular Control System: DNA, RNA, and Protein Synthesis

A. Proteins are the end result of expression of most genes.

1. Genes are expressed selectively according to the type and needs of the cell.
   a. Each **gene** encodes information for specifying a protein or type of RNA.
   b. Gene expression is the principal factor determining structural, functional, and behavioral characteristics of cells.

2. The sequence of bases in the DNA is colinear with the sequence of amino acids in the polypeptide chain.
   a. Information in DNA is transcribed into **mRNA**, which is translated into a polypeptide.
   b. Each of the 20 amino acids is specified by a three, a triplet of bases in the DNA, and a complementary codon in mRNA.
   c. The **genetic code** is consistent, universal, and degenerate.

B. New nucleic acids are synthesized by copying a template.

1. **DNA replication** occurs during interphase and results in two daughter molecules of DNA identical to the parent.
   a. Replication is **semiconservative** and is catalyzed by **DNA polymerase**, which reads the template and follows base pair rules to incorporate the correct nucleotide in the growing DNA strand.
   b. Because the strands in the parent are **antiparallel** and the polymerase adds only nucleotides in the 5′ → 3′ direction, one of the new strands grows as a continuous piece and the other is synthesized in pieces, which are later enzymatically joined together.
   c. The polymerase also proofreads the new chain as it is synthesized and makes corrections.

2. Three kinds of RNA are transcribed from genes: mRNA, tRNA, and rRNA.
   a. RNA differs from DNA in the type of sugar, the types of bases, its strandedness,

and shape in the case of tRNA.

b. RNA polymerase begins reading the **DNA template** in the 3′ → 5′ direction at the promoter and ends at the terminator site.

c. Most RNA molecules undergo post-transcription modifications to form the functional RNA.

C. New proteins are synthesized by translating mRNA.

1. All three types of RNA are involved in pro tein synthesis.
   a. **mRNA** carries to the ribosome the codons specifying amino acids to be used in the polypeptide chain.
   b. **rRNA** complexes with proteins to form the ribosome, the site for protein synthesis.
   c. **tRNA** has an **anticodon** complementary to the codon in the mRNA and an attachment site for a specific amino acid.
   d. **Amino acyl-tRNA synthetase** couples the tRNA with its correct amino acid:

$$\text{amino acid} + \text{tRNA} + \text{ATP} \rightarrow \text{aminoacyl} - \text{tRNA} + \text{AMP}$$

2. Initiation of protein synthesis requires the formation of the **ribosome**.
   a. During **initiation**, a complex forms between the mRNA, ribosome, and tRNA carrying methionine, the first amino acid specified by the **initiator codon, AUG**.
   b. A second tRNA carrying the amino acid specified by the second codon in the mRNA joins the complex by binding to a special site in the ribosome.

3. As the mRNA is read, the polypeptide chain is elongated.
   a. The cycle of **elongation** is complex and begins with a reaction catalyzed by **peptidyl transferase** : (1) A peptide bond forms between the first two amino acids; (2) the second tRNA changes position in the ribosome; (3) the first tRNA leaves and a third tRNA carrying the amino acid specified by the third codon of the mRNA enters the complex.
   b. The cycle of elongation continues until a terminator codon is exposed on the mRNA.
   c. More than one ribosome may translate the mRNA.

D. Gene expression is regulated at several different steps.

1. Regulation of gene expression occurs primarily at the level of **transcription**, but may also occur at the levels of post-transcriptional modification, translation, and post-translational modification.

2. Genetic control mechanisms can involve hormones and special regulatory proteins.

E. Damaged DNA can be repaired.

1. DNA is sensitive to damage by environmental factors, such as UV light, X-rays, and common chemicals.

2. In **photoreactivation**, an enzyme using the energy from visible light repairs damage caused by UV light.

3. Enzymes, e.g., endonucleases, polymerase, and ligase, act sequentially to excise damaged DNA and fill in the gap with the correct bases specified by the complementary strand.

4. DNA glycosylases recognize and excise altered bases and join them to the old DNA.

II. Cell-to-Cell Signaling: Hormones, Neurotransmitters, and Other Molecules

A. Adjacent cells communicate directly via gap junctions.

1. **Gap junctions** provide for metabolic and electrical coupling between cells.

2. Small molecules may pass from one cell to another through the gas junction.

B. Distant cells use chemical signals for communication.

1. Three components of chemical signaling systems are:
   a. Extracellular signals or first messengers.
   b. Receptors in either the membrane or cytoplasm of the target cell.
   c. Intracellular signals or second messengers.

2. Extracellular signals are the first messengers.
   a. They may be **local chemical mediators,** such as neurotransmitters or hormones,

produced by specialized cells and transported via the blood to target tissues.
   b. Chemical signals can be classified as water soluble or lipid soluble, which differ in the timing and duration of the responses they elicit.
   c. Prostaglandins and histamines are local chemical mediators.

3. Receptors act as mediators between first and second messengers.
   a. A **receptor**, usually a protein, binds the first messenger with high specificity and affinity.
   b. Binding of the **ligand** to the receptor triggers a sequence of events that leads to a cellular response. The receptor acts as a **transducer**.
   c. The cellular location and mechanism of action of hormone receptors vary with the type of hormone.
   d. The number of receptors in the plasma membrane for a specific agonist varies depending on the quantity of the agonist in the blood.

4. Two types of calcium-binding proteins are true receptors with no enzymatic activity: calmodulin and calcium ion regulated enzymes, such as C-kinase.
   a. Binding of calcium to the first type alters the protein's association with other proteins.
   b. Binding of calcium to the second type activates the enzyme.

5. Intracellular signals are second messengers.
   a. Binding of a first messenger usually alters the receptor's conformation. This generates a **second messenger**, which initiates intracellular responses that modify the **target cell**.
   b. Cyclic AMP, calcium ions, inositol trisphosphate, and diacylglycerol act as second messengers.
   c. They differ in their mechanism of action to produce an intracellular response.
   d. Some activate certain enzymes that catalyze the phosphorylation of certain proteins, which leads to the physiological response of the cell to the first messenger.
   e. In the absence of second messenger, the phosphorylation is readily reversed by an enzyme-mediated reaction.
   f. In the absence of extracellular signals, the concentration of second messengers is low.

Key Terms

agonist	ligand
antagonist ribosomes	local chemical mediator
anticodon	messenger RNA (mRNA)
codon	metabolic coupling
cyclic AMP	ribosomal RNA (rRNA)
DNA ligase	receptor
DNA polymerase	RNA polymerase
DNA repair	RNA splicing
electrical coupling	second messenger
first messenger	transcription
gap junction	transfer RNA (tRNA)
genetic code	translation
genome	triplet
lagging strand	
leading strand	

Testing Your Knowledge

I. Matching. Select the correct word from the Key Terms.

________ 1. sequence of three bases in DNA specifying amino acid

________ 2. sequence of three bases in tRNA

________ 3. process of protein synthesis

________ 4. molecule that binds to a receptor and elicits a response

________ 5. enzyme for transcription

________ 6. enzyme for replicating DNA

________ 7. enzyme joining together DNA pieces

________ 8. process of RNA synthesis

________ 9. carries amino acid to growing polypeptide chain

________ 10. carrier information for specifying amino acid chain to ribosomes

________ 11. second messenger

________ 12. occurs via gap junction

________ 13. located in cell membrane, binds to ligand

_________ 14. sequence of three bases in mRNA specifying amino acid

_________ 15. composed of Okazaki fragments

II. Dichotomous. Circle the correct choice from the paired terms in parentheses.

1. The (codon/anticodon) is associated with the mRNA molecule.
2. RNA is made off the DNA template in the process of (transcription/translation).
3. In DNA adenine forms a base pair with (uracil/thymine).
4. In the processing of mRNA, sequences corresponding to (exons/introns) in the DNA are excised.
5. In the initiation complex, the tRNA carrying the first amino acid binds to the (A site/P site) on the ribosome.
6. Insulin would bind to a (plasma membrane/cytoplasmic) receptor.
7. Testosterone would bind to a (plasma membrane/cytoplasmic) receptor.
8. An (agonist/antagonist) evokes a cellular response.
9. Cyclic AMP is an example of a (first/second) messenger.
10. The ion most commonly behaving as a second messenger is (calcium/sodium).

III. True/False. In the space to the left of the number write T for true or F for false.

_______ 1. Each of the 64 codons specifies an amino acid.

_______ 2. Proofreading of the DNA molecule is done by DNA polymerase.

_______ 3. An anticodon is found in tRNA.

_______ 4. Each mRNA is translated by only one ribosome at a time.

_______ 5. Mutations in DNA cannot be repaired.

_______ 6. Neurotransmitters bind to receptors located in the cytosol.

_______ 7. A daughter molecule of DNA has one of the parental strands and one new strand.

_______ 8. Introns are non-coding sequences in the DNA.

_______ 9. The number of receptors is proportional to the number of ligands in the blood.

_______ 10. Neurotransmitters pass between cells through gap junctions.

IV. Sentence Completion. Write your response in the space provided.

1. AUG is the codon for _______________.
2. A tRNA and its amino acid are joined together by the action of _______________.
3. Metabolic and electrical coupling occur through _______________ in the plasma membranes of adjacent cells.
4. A _______________ soluble ligand would bind to a receptor on the plasma membrane.
5. There are _____________ different amino acids and _____________ different codons.
6. The anticodons in the _______________ are complementary to the codons in the _______________.
7. A cell that responds to insulin would have _________ on its plasma membrane.
8. Clusters of ribosomes are called _______________.
9. Before an amino acid may become part of a growing polypeptide chain, it must be joined to a _______________.
10. A _________________ allows small molecules to pass between cells.

V. Multiple Choice. Circle the letter of the best response.

1. Which of the following is not found in DNA?
   a. uracil
   b. adenine
   c. thymine
   d. guanine
   e. cytosine

2. Adenine is paired with thymine in DNA by:
   a. hydrogen bonds
   b. covalent bonds
   c. disulfide bridges
   d. a double bond
   e. none of the above

3. Protein synthesis is known as:
   a. transcription
   b. replication
   c. translation
   d. elongation
   e. excision

4. For a DNA strand 3′A A T C G A G T C 5′, the complementary RNA is:
   a. 3′ U U A G C U C A G 5′
   b. 3′ A A T C G A G T C 5′
   c. 5′ T T A G C T C A G 3′
   d. 5′ U U A G C U C T G 3′
   e. none of the above

5. Which of the following occurs in post-transcriptional modification of a pre-mRNA?
   a. addition of a 7-methylguanosine cap
   b. addition of a poly A tail
   c. excision of sequences corresponding to introns
   d. joining of sequences corresponding to exons
   e. all of the above

6. Two DNA fragments are joined together by:
   a. ligase
   b. DNA polymerase
   c. topoisomerase
   d. synthetase
   e. DNA glycosylase

7. Which of the following is not true for receptors?
   a. They are usually proteins.
   b. They show specificity for the ligand.
   c. They bind with high affinity to the ligand.
   d. Their number remains constant throughout the life of the cell.
   e. none of the above

8. Which of the following is true for calcium as a second messenger?
   a. The resting cellular level of Ca^{2+} is > 1 μM.
   b. Ca^{2+} is actively transported into the cell when the ligand binds to the receptor.
   c. It is less versatile than cyclic AMP as a second messenger.
   d. Ca^{2+} may bind directly to phospholipid-dependent protein kinase.
   e. The physiological response to Ca^{2+} is the same in all cell types.

9. In general, second messengers:
   a. are present in very low concentrations in a resting cell
   b. do not turn over rapidly
   c. remain in the cytosol long after the ligand is released from the receptor
   d. always produce a physiological response through protein phosphorylation
   e. all of the above

10. All of the following are second messengers except:
    a. cyclic AMP
    b. Ca^{2+}
    c. inositol trisphosphate
    d. diacylglycerol
    e. phospholipase C

VI. Complete the Table.

	Type of Sugar	Bases (A, T, C, G, U)	# of Strands	Function
DNA				
mRNA				
tRNA				
rRNA				

VII. Complete the Table.

	?	?	?	?	?
DNA			ACA		
mRNA		UCG			UAG
tRNA				CAU	
amino acid	met				

Chapter 6

Energy and Cellular Metabolism

Chapter Learning Objectives

1. Describe the general energy transactions that are carried on by cells.
2. Compare and contrast photosynthesis and respiration as well as endergonic and exergonic reactions in general.
3. Provide an overview of the catabolism of the major organic compounds. Consider the roles of electron acceptors, coenzymes, and oxygen in the oxidation-reduction reactions occurring in the cell.
4. Describe the process of glycolysis in detail. Consider the reactants, products, and num bers of ATP and NADH molecules produced.
5. Describe the citric acid cycle in detail. For the reaction in which pyruvate is degraded to acetyl CoA, indicate the reactants, products, and numbers of ATP and NADH molecules produced.
6. Describe the electron transport chain in detail. Consider the movement of electrons down the chain and the role of cytochromes.
7. Describe the chemiosmotic mechanism for the production of ATP and account for each of the 36 molecules of ATP produced from the complete oxidation of one molecule of glucose.
8. Trace the pentose phosphate pathway and describe how it fits into cellular metabolism. Consider the reactants, products, and importance of this pathway.
9. Present the metabolism of fats and relate it to cellular metabolism.
10. Show how proteins enter into cellular metabolism.

Expanded Chapter Outline

I. Basic Principles of Energy Transformation in the Cell

A. Many factors influence the rate and duration of chemical reactions in cells.

1. Reactions with a negative ΔG are **exergonic**, whereas those with a positive ΔG are **endergonic**.

2. Reactions are driven to completion by removal of the end product.
   a. Reverse reactions occur depending on the needs of the cell.
   b. Many reactions take place with a small change in free energy.
   c. No reaction occurs independently or in isolation.

3. In an **oxidation** reaction, an H atom or electrons are given up; in a **reduction** reaction, an H atom or electrons are gained.
   a. Generally, compounds with many H atoms are high energy; compounds with few are low energy.
   b. Photosynthesis is a reduction process, and cellular respiration is an oxidation process.
   c. H atoms and electrons are transferred first to coenzymes such as **NAD^+** then to oxygen via the **electron transport chain**.
   d. NAD^+ has a negative **redox potential** and low affinity for electrons in contrast to oxygen.

4. Phosphate compounds store high energy.
   a. When one of the phosphate groups from ATP, a high-energy molecule, is transferred to an acceptor, a large amount of free energy is released.
   b. In the cell, the free energy change (ΔG) is minus 12 Kcal/mol for the reaction $ATP \rightarrow ADP + P_i$.
   c. Compounds with a less negative ΔG have a greater affinity for the phosphate group and serve as acceptors for groups donated by ATP.
   d. ATP is produced by substrate level and oxidative phosphorylation.

5. Enzymes play a central role in cellular metabolism.

a. Some enzymes require **coenzymes**, such as coenzyme A and NAD, for their activity.
b. Cells control enzyme-catalyzed reactions by modification of the enzyme; by **allosteric inhibition**, a reversible process; or by increasing or decreasing synthesis of the enzyme.
c. Reduction of NAD^+ is expressed by the reaction: $NAD^+ + 2H^+ + 2e^- \rightarrow NADH + H^+$

II. An Overview of Catabolism: Dietary Proteins, Carbohydrates, and Fats

~ In the intestine, proteins, carbohydrates, and fats are broken down into amino acids, sugars, fatty acids, and glycerol.

~ In the cell, amino acids, monosaccharides, and fatty acids are degraded further by the processes of deamination, glycolysis, and fatty acid oxidation, all of which are regulated and facilitated by enzymes.

~ Molecules from the degradation of amino acids, glucose, and fatty acids are converted to acetyl groups in the mitochondrial matrix.

~ Acetyl groups are coupled to coenzyme A to form acetyl CoA, which is converted to CO_2 via the citric acid cycle. High-energy electrons released from this conversion are temporarily trapped by NAD and FAD, which feed the electrons to the respiratory chain.

~ Redox reactions occur as electrons pass down the chain to oxygen; released energy is used to synthesize ATP in a process called oxidative phosphorylation.

III. The Metabolism of Carbohydrates: The Oxidation of Glucose

A. *Glycolysis is the conversion of glucose to pyruvate.*

1. Glucose is degraded into two molecules of pyruvic acid.
   a. Enough net free energy is released to make two molecules of ATP by substrate level phosphorylation.
   b. Two molecules of NADH are also produced.
2. Each of the ten reaction steps is enzyme-catalyzed and takes place in the cytosol.
3. The first stage of glycolysis uses two ATP molecules for phosphorylating and converting glucose → fructose 1,6-bisphosphate.
4. In the second stage, two molecules of **NADH** and four ATP molecules are produced as glucose is metabolized to pyruvate.
   a. Glycolysis, an **anaerobic** process, yields less than 3% of the maximum amount of free energy stored in glucose.
   b. Under anaerobic conditions, pyruvate → lactic acid.

B. *The citric acid cycle provides reducing power as NADH and $FADH_2$.*

1. Pyruvic acid enters the mitochondrial matrix, where it is coupled to coenzyme A in preparation for processing via the **citric acid cycle**.
   a. In the reaction of **pyruvate → acetyl CoA**, molecules of CO_2 and NADH are produced.
   b. Degradation of acetyl to two molecules of CO_2 occurs in the citric acid cycle.
2. For each acetyl group entering the cycle, three molecules of NADH, one molecule of FADH, and high-energy phosphate GTP are generated.
3. Each step in the citric acid cycle is enzyme-catalyzed and occurs in the mitochondrial matrix.
4. The cycle may be controlled by the concentrations of ATP and ADP acting as allosteric inhibitors and activators, respectively, as well as other mechanisms.

C. *Electron transfer is coupled to synthesis of ATP.*

1. The oxidation of a molecule of **glucose → CO_2** yields 10 molecules of NADH, 2 mole cules of $FADH_2$, and 4 molecules of ATP. Thirty-two molecules of ATP are produced by oxidative phosphorylation.
2. **Cytochromes** are electron carriers.
   a. NADH and $FADH_2$ have very negative redox potentials and readily donate electrons to the electron carrier proteins in the respiratory chain.

b. Redox potentials of the carriers increase sequentially from the beginning to the end of the chain, which facilitates the transfer of electrons along the cytochromes to oxygen.

3. At three sites in the electron transport chain, sufficient energy is released to synthesize a molecule of ATP.
   a. Three molecules of ATP are produced for each molecule of NADH oxidized.
   b. For each $FADH_2$ that is oxidized, two molecules of ATP are synthesized.

4. ATP is synthesized by the chemiosmotic mechanism.
   a. As electrons flow down the respiratory chain, protons are pumped from the matrix to the intermembrane space, thus setting up an **electrochemical gradient** across the inner mitochondrial membrane.
   b. As protons flow down the gradient and through special channels in the membrane, the enzyme **ATP synthetase** converts ADP → ATP.
   c. ATP synthetase transforms the energy of proton gradient into chemical energy of ATP in a process known as the **chemiosmotic mechanism.**

D. *The pentose phosphate pathway is an alternative for glucose.*

1. This pathway generates pentose sugars and NADPH.

2. NADPH is used in the oxidation-reduction reactions occurring in various biosynthetic pathways, such as the synthesis of fatty acids and steroids.

IV. Metabolism of Fats and Proteins

A. *Catabolism of fats occurs in the mitochondrial matrix.*

1. In a series of complex reactions, long-chain fatty acid is converted into acetyl CoA molecules, which enter the citric acid cycle.

2. Fatty acids are synthesized from acetyl CoA by a pathway different from that used for their oxidation.

B. *Deaminated amino acids enter the pathway of glucose oxidation.*

C. *Acetyl coenzyme A is an intermediary between catabolism and biosynthesis.*

1. Most of the intermediate compounds produced by catabolism are fed into the citric acid cycle as acetyl CoA.

2. Acetyl CoA is also a precursor used in the synthesis of fatty acids and glucose.

Key Terms

acetyl
aerobic
allosteric inhibition
anaerobic
ATP synthetase
biosynthesis
catabolism
cellular respiration
chemiosmotic synthesis
citric acid cycle
coenzyme A
cytochrome
deamination
electron transport chain
FAD^+, $FADH_2$
gluconeogenesis
glycogenolysis
glycolysis
mitochondria
mitochondrial matrix
NAD^+, NADH
NADP, NADPH
oxaloacetate
oxidation
oxidative phosphorylation
photosynthesis
redox potential
reduction
substrate level phosphorylation

Testing Your Knowledge

I. Matching. Select the correct word from the Key Terms

________ 1. term analogous to end product inhibition

________ 2. oxidation of glucose to pyruvic acid

________ 3. reaction by which pyruvic acid is degraded to CO_2

________ 4. reaction in which electrons and protons are gained

________ 5. reaction in which electrons and protons are lost

________ 6. ATP production occurring in cytoplasm

________ 7. measure of affinity of compound for electrons

_______ 8. location of reactions of fatty acid oxidation

_______ 9. electron carrier in respiratory chain

_______ 10. molecule regenerated in citric acid cycle

_______ 11. removal of amino group from amino acid

_______ 12. common intermediary in catabolic and biosynthetic pathways

_______ 13. electron and proton carrier produced by pentose phosphate pathway

_______ 14. channel protein through which protons flow

_______ 15. $C_6H_{12}O_6 + 6O_2 \rightarrow 6CO_2 + 6H_2O +$ energy

II. Dichotomous. Circle the correct choice from the paired terms in parentheses.

1. The cell extracts energy from its environment in a series of reactions known as (catabolism/anabolism).
2. Delta G is (negative/positive) in an endergonic reaction.
3. The reduced form of NAD is (NAD^+/NADH).
4. A molecule with a low or negative redox potential has a (high/low) affinity for electrons.
5. Oxidative phosphorylation occurs in the (cytoplasm/mitochondria).
6. If the cellular ATP supply is low, the citric acid cycle is (slowed/accelerated).
7. In the reaction $1/2\ O_2 + NADH + H^+ \rightarrow H_2O + NAD^+$, oxygen is (reduced/oxidized).
8. The electron acceptor for the pentose phosphate pathway is ($NADP^+$/NAD^+).
9. When oxidized the energy from one molecule of glucose generates (more/less) ATP than that from one molecule of fatty acid.
10. Nonessential amino acids (can/cannot) be made in the body.

III. True/False. In the space to the left of the number write T for true or F for false.

_______ 1. ATP is a long-term energy storage molecule.

_______ 2. End product inhibition is also known as allosteric inhibition.

_______ 3. Glycolysis occurs in the cytosol.

_______ 4. Glycolysis occurs only under aerobic conditions.

_______ 5. When ATP acts as an allosteric inhibitor of citrate synthetase, the cycle is slowed.

_______ 6. If the supply of ATP is adequate, the cycle is slowed.

_______ 7. The flow of protons down the gradient is coupled to the synthesis of ATP.

_______ 8. Proteins provide more energy per unit weight than fats.

_______ 9. The electron transport chain cytochromes are located in the inner mitochondrial membrane.

_______ 10. The breakdown of amino acids begins with deamination.

IV. Sentence Completion. Write your response in the space provided.

1. _______________ uses CO_2 and produces O_2 as a byproduct, but _______________ uses O_2 and produces CO_2 as a byproduct.
2. In _______________ reactions, electrons are lost, and in _______________ reactions, they are gained.
3. Electrons will be transferred from a molecule with a _______________ redox potential to one with a _______________ redox potential.
4. The reaction $ATP \rightarrow ADP + P_i$ has a _______________ ΔG.

5. Most of the breakdown products of amino acids, glucose, and fatty acids are converted to ______________ groups in the mitochondrial matrix.

6. Pyruvate is converted to acetylCoA in the ______________.

7. The complete oxidation of glucose generates __________ molecules of NADH and __________ molecules of $FADH_2$.

8. For each molecule of NADH oxidized, ________ molecules of ATP are synthesized.

9. To set up the proton gradient, protons are pumped from the ______________ to the ______________.

10. Fatty acid oxidation occurs in the ______________, and fatty acid synthesis occurs in the ______________.

V. Multiple Choice. Circle the letter of the best response.

1. In the reaction $AB \leftrightarrow A + B$, if B is removed, the:
   a. reaction will reach equilibrium
   b. forward reaction is favored
   c. reverse reaction is favored
   d. concentration of AB will increase
   e. concentration of A will increase

2. In the reaction $(CH_2O)n + O_2 \rightarrow CO_2 + H_2O$ + energy:
   a. O_2 is oxidized
   b. $(CH_2O)n$ is reduced
   c. O_2 is reduced
   d. $(CH_2O)n$ is oxidized
   e. c and d are correct

3. All of the following are coenzymes except:
   a. NAD
   b. FAD
   c. NADP
   d. COA
   e. GTP

4. Which of the following is true for glycolysis?
   a. Glucose is converted to acetylCoA.
   b. Glucose is completely oxidized to CO_2 and H_2O.
   c. It occurs in the mitochondrial matrix.
   d. It is an example of anabolism.
   e. Two ATPs are used and four are produced.

5. Which of the following is not true for the citric acid cycle?
   a. Oxaloacetate is regenerated.
   b. The cycle cannot operate under anaerobic conditions.
   c. NAD and FAD are reduced.
   d. 36 molecules of ATP are produced directly.
   e. If ATP supply is adequate, the cycle is slowed.

6. The electrical proton gradient for chemiosmotic synthesis of ATP is created across the:
   a. inner mitochondrial membrane
   b. outer mitochondrial membrane
   c. intermembrane space
   d. mitochondrial matrix
   e. plasma membrane

7. The final hydrogen acceptor in cells is:
   a. oxygen
   b. NAD^+
   c. FAD^+
   d. $NADP^+$
   e. H_2O

8. A radioactively labeled carbon atom in an amino acid might be found in:
   a. pyruvate
   b. acetylCoA
   c. intermediates of the citric acid cycle
   d. a and b are correct
   e. a, b and c are correct

9. Which of the following is true for pyruvate?
   a. Under anaerobic conditions, it is converted to acetylCoA in the cytosol.
   b. When it is converted to acetylCoA, CO_2 is released.
   c. It enters the citric acid cycle directly
   d. For each molecule of glucose oxidized, one molecule of pyruvate is produced
   e. none of the above

10. Cyanide interferes with the last redox reaction in the electron transport chain. In the presence of cyanide, ATP production would:
    a. be unaffected
    b. accelerate
    c. slow down
    d. stop

VI. Fill in the Table

Reaction Series	Product(s)	No. of ATP Produced	No. of NADH Produced	Location of Reaction
glycolysis				
Pyruvic acid to acetylCoA				
oxidative phosphorylation				
fatty acid oxidation of palmitic acid				

Chapter 7

Functional Organization of the Nervous System

Chapter Learning Objectives

1. Give an overview of the nervous system, its structure, and its components.
2. Describe the resting potential of a nerve cell.
3. Indicate what happens to produce an action potential.
4. Compare electrical and chemical synapses.
5. Describe axonal transport as to the role of cytoskeletal components, substances moved, direction of movement, and rate.
6. Describe how ion channels work.
7. Explain how equilibrium potential affects nerve cells.
8. Diagram the components of an action potential.
9. Describe what happens at a synapse.
10. List the major neurotransmitters; describe their chemical structure and production, how they bind to receptors, their effect on membrane channels, and their fate after uncoupling from the receptor.

Expanded Chapter Outline

I. Overview: The Central and Peripheral Nervous Systems

II. Cells of the Nervous System

A. Glia are multifunctional cells.

1. **Oligodendroglia** form myelin sheaths around axons to prevent the passage of ions through the axonal membrane.

2. The functions of **Schwann cells** include:
   a. Ensheathing axons in the peripheral nervous system.
   b. Axonal regeneration.

3. **Astroglia** exhibit the greatest diversity of function among glial cells in the central nervous system.
   a. Fibrous and protoplasmic astrocytes are filamentous and form "glial end feet" on blood vessels.
   b. Reactive astrocytes increase in number after brain injury.
   c. Astrocytes regulate the concentrations of potassium ions and neurotransmitters during synaptic transmittance.
   d. Astrocytes are also believed to be involved in the maintenance of interstitial pH.

4. **Microglia** are found near blood vessels and function to remove cellular debris.

B. Neurons, the conductors of nervous impulses, are specialized cells.

1. Neurons have four functionally distinct regions: the soma, dendrites, axon, and axon terminal.

2. They are classified structurally as bipolar, multipolar, or unipolar and functionally as sensory, motor, or interneurons.

3. Differences from other cells in the body include:
   a. Nuclear chromatin in the form of dispersed chromatin because of its high transcriptional activity.
   b. Rough ER and Golgi bodies found predominantly in the soma.
   c. Smooth ER storing calcium ions and playing a role in maintaining intracellular calcium levels.
   d. Mitochondria concentrated in axonal terminals.

4. Ion channels in membranes may be classified as passive, chemically activated, or voltage activated.

5. A neuronal cytoskeleton in neuronal processes is a complex of filaments and transport proteins.
   a. **Microfilaments** are found in dendritic spines and growth cones.

b. **Neurofilaments,** found in axons and dendrites, provide structural rigidity.
c. **Microtubules** play a role in the rapid transport of molecules in axons and dendrites.

6. Axonal and dendritic transport mechanisms include:
   a. **Anterograde** and **retrograde movements** mediated by microtubule-associated proteins kinesin and dynein, respectively.
   b. **Slow and fast axonal transports** that differ in rate; type of material being transported; and the requirement for calcium ions, glucose, and ATP.
   c. **Dendritic anterograde transport** that requires ATP and is involved in movement of ribosomes and RNA.

7. Nerve fiber development and regeneration occur from a growth cone.
   a. Axonal elongation occurs at the growth cone, which is present during normal development and regeneration.
   b. Axonal growth in a specific direction is determined largely by interactions between the growing axons, trophic factors secreted by target cells, and cell adhesion molecules (CAMs).

III. The Membrane Potential and the Action Potential

A. Electrical potentials are produced across the membrane of resting cells.

1. The electrical potential difference across the membrane, Em, is about minus 60 mV.

2. The movement of ions across their channels results from a difference in the electrical charge or concentration of ions across the plasma membrane.

3. The conductance of an ion through a channel is affected by the size of the ion relative to that of the channel, the ion's charge, and state of hydration.

4. The intracellular concentration of K^+ is greater than that in the extracellular compartment, while the reverse is true for Na^+.

5. **Ohm's law, I = G times E,** represents the relationship between electrical potential (E), ionic current (I), and conductance (G).

6. The **equilibrium potential** can be expressed by the **Nernst equation:**

E_{ion} = RT/zF times $\log_e$ $(ion)_o/(ion)_i$
where R = gas constant, T = absolute temperature, z = ion valence, F = Faraday constant, and E = equilibrium potential.

7. The factors affecting resting membrane potential may be expressed by the **Goldman equation.**

B. The action potential is a temporary change in the membrane potential.

1. The **action potential** is a temporary change from a negative resting state to a positive value.

2. The voltage changes in the action potential result from the opening and closing of voltage-sensitive ion channels that control the influx of Na^+ and efflux of K^+.

3. The action potential, an "all-or-none phenomenon," is initiated once the threshold value of the membrane potential has been reached.

4. During the **absolute refractory period,** the axon is unable to generate a second action potential.

5. The conduction velocity is proportional to the diameter of the axon and can occur rapidly by saltatory conduction in myelinated neurons.

IV. Communication Between Nerve Cells: Synaptic Transmission

A. Electrical synapses mediate communication between adjacent cells.

B. Chemical synapses mediate communication between distant cells by transmitter-receptor interactions.

1. The process of synaptic transmission involves four phases:
   a. Synthesis of transmitter.
   b. Storage and release of transmitter.
   c. Interaction of the transmitter with postsynaptic receptor.

d. Termination of synaptic transmission.

2. Calcium ions are required for the release of transmitters from the presynaptic cell.

3. **Excitatory post-synaptic potentials (EPSPs)** depolarize the membrane potential of post-synaptic neurons by opening Na^+ channels.

4. **Inhibitory post-synaptic potentials (IPSPs)** hyperpolarize the membrane potential of post-synaptic neurons by opening K^+ channels.

5. **Serotonin**, derived from tryptophan, interacts with several types of receptors whose activation may generate IPSPs or EPSPs, depending on the receptor.

6. **Glutamate**, an amino acid, binds to three types of receptors whose activation produces EPSPs by opening either Na^+ and K^+ ion channels or increasing $Ca2^+$ conductance.

7. **GABA**, synthesized from glutamate, binds to two types of receptors, A and B.
   a. Receptor A is a ligand-gated Cl^- channel whose activation leads to an IPSP.
   b. Activation of receptor B leads to an IPSP via increased K^+ conductance.

8. **Hyperpolarization** and depolarization may result from the transmitter-activated closing of chemically activated ion channels.

9. Termination of synaptic transmission occurs when the transmitter is removed from the synaptic cleft by reuptake and recycling in the presynaptic cell or by degradation.

10. In a **discrete synapse**, the neurotransmitter is released from active zones in the presynaptic terminal into a small synaptic cleft. In a **diffuse synapse**, transmitter release is not limited to specific active zones, and the cleft can be as large as 150 nm.

V. Neurotransmitters and Associated Receptors

A. Low-molecular-weight transmitters are produced at axon terminals.

1. Low-molecular-weight transmitters, such as acetylcholine, biogenic amines, and amino acids, are synthesized within the presynaptic terminal.

2. The two types of acetylcholine receptors, nicotinic and muscarinic, differ in their effect on the channel, their structure, and their function.

B. High-molecular-weight neuropeptides are pro duced in the soma.

1. Neuropeptides are synthesized and packaged into vesicles in the soma then carried to the terminal by fast axonal transport.

2. Neuropeptides are classified into one of several families of peptides based on their amino acid sequence and function.

3. Dopamine binds to one of two types of receptors, D1 and D2, that are coupled to G_s and G_i proteins, respectively; binding affects the activity of adenylate cyclase.

4. The synthesis of norepinephrine, a catecholamine derived from dopamine, is regulated through modulation of the enzymes tyrosine hydroxylase and dopamine beta-hydroxylase.

VI. Neuronal Integration: Temporal and Spatial Summation

A. The action potential discharge by a nerve cell is based on the summation of excitatory and inhibitory post-synaptic potentials.

B. These potentials occur in time over the space of the neuron's receptive surface.

Key Terms

absolute refractory period
acetylcholine
activating state
axonal transport
biogenic amine
depolarization
diffuse synapse neuropeptide
discrete synapse
efferent (motor) neuron
electrical synapse
excitatory postsynaptic potential (EPSP)
glial cell
growth cone
hyperpolarization
inhibitory post-synaptic potential (IPSP)
interneuron
ion channel
muscarinic acetylcholine receptor
neurotransmitter
nicotinic acetylcholine receptor
nodes of Ranvier
post-synaptic neuron
pre-synaptic neuron
receptor

regeneration
relative refractory period
repolarization
saltatory conduction
serotonin
soma
summation (spatial, temporal)
synaptic cleft
synaptic vesicle

Testing Your Knowledge

I. Matching. Select the correct term from the Key Terms.

_________ 1. describes state of sodium channel at minus 45 mV

_________ 2. period during which axon cannot generate a second action potential

_________ 3. movement of action potential along nodes of Ranvier

_________ 4. pathway for direct conduction of ions between cells

_________ 5. post-synaptic potential depolarizing membrane

_________ 6. form blood-brain barrier

_________ 7. post-synaptic potential hyperpolarizing membrane

_________ 8. metabolic center of a neuron

_________ 9. transmits information about stimulus to central nervous system

_________ 10. transmitter synthesized in synaptic terminal and released in neuromuscular synapse

_________ 11. contain neurotransmitter

_________ 12. synapse in which transmitter released from active zones

_________ 13. low molecular weight transmitter synthesized from tyrosine

_________ 14. transmitter synthesized in soma

_________ 15. describes situation in which two or more synaptic outputs activated together cause electrical potential to become positive.

II. Dichotomous. Circle the correct choice from the paired terms in parentheses.

1. (Sensory/Motor) neurons transmit information from the central nervous system to muscle.
2. (Neurons/Glial cells) transmit information.
3. The (axons/dendrites) receive information from other neurons.
4. The (rough/smooth) ER binds calcium to maintain a low cytoplasmic concentration.
5. Conductance through a channel is (inversely/directly) related to diameter.
6. The concentration of (Na^+/K^+) is greater in the cytoplasm than in extracellular fluid.
7. The equilibrium potential for an ion is the potential at which efflux (equals/is greater than/is less than) influx.
8. During the (absolute/relative) refractory period, the axon is incapable of generalizing an action potential.
9. Ions, cAMP, and small molecules move directly between adjacent neurons through (chemical/electrical) synapses.
10. Hypothetically, an ion with an equilibrium potential greater than +55 mV would be (more/less) attracted into the cell compared with Na^+.

III. True/False. In the space to the left of the number write T for true or F for false.

_________ 1. At a synapse, neurons communicate via gap junctions.

_________ 2. Interneurons are found in the peripheral nervous system.

_________ 3. Most protein synthesis in a neuron occurs in the axon.

_________ 4. In a resting neuron, chromatin is visible as chromosomes.

_________ 5. Cell-adhesive molecules and trophic factors are important for neuronal growth.

_______ 6. Membrane conductance is the same for all regions of a neuron.

_______ 7. The velocity of the action potential depends on its threshold level.

_______ 8. The post-synaptic cell is always a neuron.

_______ 9. There is a direct relationship between the amount of Ca^{2+} with the axonal terminal and the amount of neurotransmitter released.

_______ 10. Binding of norepinephrine to alpha-2 receptors in pre-synaptic terminals inhibits the release of norepinephrine.

IV. Sentence Completion. Write your response in the space provided.

1. The somatic and autonomic nervous systems are components of the _______________ nervous system.

2. A _____________ neuron has one axon and one dendrite on opposite sides of the soma.

3. Mitochondria are highly concentrated in ___________________.

4. __________ axonal transport requires ATP.

5. _______________ ion channels are located predominantly on dendrites.

6. A change of membrane potential from −60 mV to −45 mV occurs during __________.

7. At the peak of the action potential, the membrane is more permeable to _________ than _________ in contrast to permeability during the resting stage.

8. The ___________ the distance between the nodes of Ranvier, the greater the velocity of propagation.

9. An influx of _____________ causes fusion of synaptic vesicles with the membrane and exocytosis.

10. The effect of the binding of ACh to _______ _______ receptors is mediated through G proteins.

V. Multiple Choice. Circle the letter of the best response.

1. Which of the following is true for oligodendroglia, but no other glial cells?
   a. They form regeneration tubes.
   b. One oligodendroglia wraps around one axon.
   c. They remove degenerating debris by phagocytosis.
   d. They regulate the concentration of K^+ in the extracellular space.
   e. none of the above

2. Which is not a function of astrocytes?
   a. regulation of extracellular K^+ concentration
   b. regulation of neurotransmitter concentration at synapse
   c. regeneration of axons
   d. removal of debris by phagocytosis
   e. maintenance of interstitial pH

3. Which of the following is true for fast axonal and not slow axonal transport?
   a. It requires calcium ions.
   b. It always occurs in a retrograde direction.
   c. The rate of transport is about 1 mm per day.
   d. Neurofilaments and microtubules are transported.
   e. Glycolysis provides energy.

4. Which of the following is not true for conductance?
   a. It is measured in units called siemens.
   b. Electrical potential is directly proportional to conductance.
   c. Ionic current is directly proportional to conductance.
   d. Hydration affects conductance through a channel.
   e. All are true.

5. Which of the following is true for a Na^+ channel and not for a K^+ channel?
   a. Sodium channels depend only on voltage, and K^+ channels depend on voltage and time.
   b. K^+ channels have inactivation and activation gates.
   c. Na^+ channels open more rapidly than K+.
   d. The Na^+ channels open during the resting state, but the K^+ channels are closed.

6. The action potential is produced by the:
   a. influx of K^+ and efflux of Na^+

b. influx of K^+ and Na^+
c. efflux of K^+ and Na^+
d. influx of Na^+ and efflux of K^+
e. influx of Na^+ and Ca^{2+} and efflux of K^+

7. Which of the following events occurs first in an action potential?
   a. activation of Na^+ gate
   b. activation of K^+ gate
   c. hyperpolarization of the membrane
   d. depolarization of the membrane
   e. regenerative process

8. Which of the following is true for excitatory post-synaptic potentials (EPSPs)?
   a. They hyperpolarize membrane potential.
   b. They result from an influx of K^+ into the cell.
   c. They result from an efflux of Na^+ from the cell.
   d. They result from an efflux of $Cl-$ from the cell.
   e. They result when the inside of the cell becomes more positive.

9. Which of the following is true for dopamine?
   a. It binds to muscarinic receptors.
   b. It is synthesized from tryptophan.
   c. It affects the activity of adenylate cyclase through G proteins.
   d. It is degraded in the synaptic cleft by DOPA dicarboxylase.
   e. It is not affected by cocaine or amphetamine.

10. Which of the following is not true for norepinephrine (NE)?
    a. It is derived indirectly from tyrosine.
    b. Its synthesis is regulated on a long-term and short-term basis.
    c. Binding of NE to beta2 receptors on presynaptic cells increases the release of NE.
    d. Binding of NE to beta2 receptors on smooth muscle cells causes an influx of Ca^{2+} and contraction.
    e. Binding of NE to beta2 receptors on smooth muscle of blood vessels causes vasoconstriction.

VI. Complete the Table.

One line has been completed.

Neurotransmitter	Type of Channel Affected	Effect on Channel	Effect on Post-Synaptic Membrane	Fate of Transmitter
e.g., transmitter	chemically activated ion channel	closes channel	depolarization	uptake into presynaptic cell
acetylchloline				
dopamine				
norepinephrine				
serotonin				

VII. Label the Diagram.

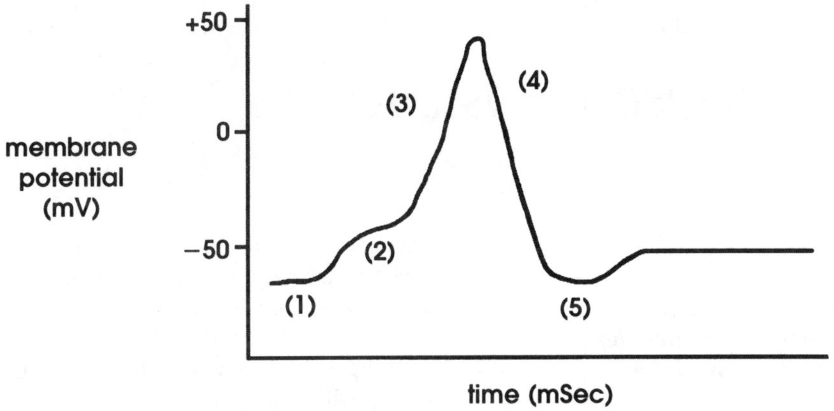

Label the diagram and describe the state of polarization of the membrane for the numbered stages.

1. ________________________________
2. ________________________________
3. ________________________________
4. ________________________________
5. ________________________________

VIII. Label the Diagram of the Neuron.

Structures to be labeled: soma, axon, cell body, anterograde direction, retrograde direction

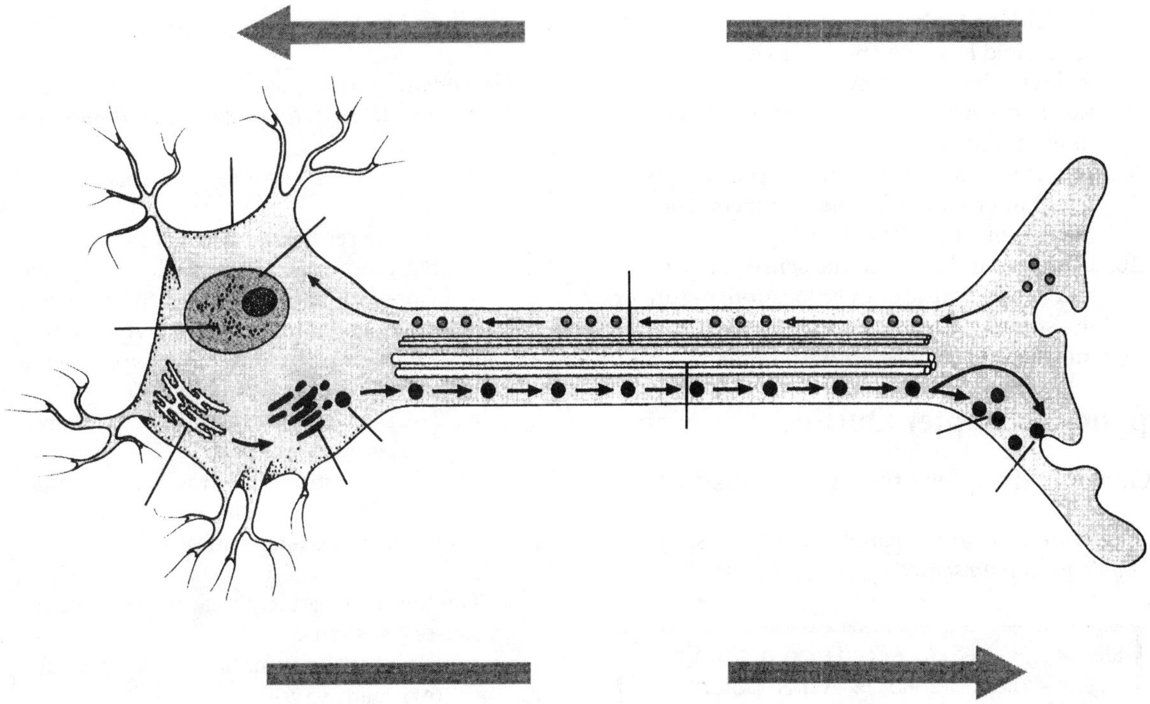

Chapter 8

Sensory Systems

Chapter Learning Objectives

1. Describe how we detect information from external stimuli; consider the common features of transduction mechanisms for all sensory systems.
2. Indicate how each type of stimulus produces differences in intensity and quality of information that travels to the central nervous system; delineate the information pathway from the receptor to the brain for a somatic sensory system.
3. Distinguish among how pressure, pain, and thermal stimuli are transmitted.
4. Present the concepts involved in a somatotopic map of the body.
5. Describe the concept of the receptive field, how the various senses have particular types of receptive fields, and how this translates into information useful for interpreting environmental information.
6. Diagram and label the parts of the eye and the retina.
7. Explain the trichromatic theory of color vision and how we are able to distinguish yellow and orange flowers.
8. Relate the structure and function of the ear to the ability to hear.
9. Describe how different noises produce different information for the brain; consider the role of hair cells in hearing.
10. Compare and contrast the senses of taste and smell; consider their location, nature of the stimulus, structure, and pathway of information from receptor to brain.

Expanded Chapter Outline

I. General Principles of Sensory Transduction

A. *The transduction mechanism for all sensory systems is represented by the sequence:*

> Stimulus → Membrane conductance change → Generator potential → Action potential

B. *The intensity of a stimulus is encoded by the frequency of action potentials in the receptor cell or neuron connected to it.*

1. Each type of sensory receptor is activated by a specific stimulus.

2. Information about the intensity of a stimulus is conveyed to the brain in coded form.
   a. The **frequency code** reflects the number of action potentials generated per unit time in a single fiber.
   b. The **population code** reflects the number of sensory neurons producing action potentials.

3. The quality of a stimulus is encoded by the pathway of transmission.

4. Nerves adjust to repetitive stimuli by mechanisms of adaptation, which occurs at the molecular level.

II. Sensory Systems and Their Processes

A. *The somatic sensory system transduces information about pain, temperature, and pressure.*

1. Different types of somatic sensory systems include:
   a. **Tactile receptors**, also called mechanoreceptors, that detect touch, pressure, and vibrations applied to the skin. Tactile receptors may be rapidly or slowly adapting.
   b. **Thermal receptors** that respond to temperatures above or below 30° C.
   c. **Nociceptors** that respond to pain and temperature above 45° C.
   d. **Proprioceptors** that transmit information about the position of the body's limbs and changes in position.

2. The following principles apply to all somatic sensory systems:
   a. Each type of somatic sensation has its own pathway.

b. Most somatic pathways cross over from one side of the brain to the other.
c. Each pathway is made of connections between clusters of cell groups called nuclei.
d. Nerve cells within each cluster are topographically organized according to the location of their sensory receptor in the surface of the body.

3. Cell bodies of sensory neurons are located in **dorsal root ganglia**.

4. The spinal cord is subdivided into functional areas and segments.
   a. Each of the 31 segments receives sensory information from particular areas of the skin called **dermatomes**.
   b. The **dorsal column** and **anterolateral pathways** transmit somatic sensory information to nerve cell nuclei in the spinal cord, brain stem, thalamus, and neocortex.

5. The **thalamus** receives most types of sensory information, including somatic, visual, auditory, and taste.

6. The **somatic sensory cortex** region of the **neocortex** is divided into discrete areas to receive somatic, visual, auditory, and gustatory sensations.
   a. The somatic sensory cortex is topographically organized to correspond to regions of the body.
   b. The body regions with the greatest density of sensory receptors are represented by a greater proportion of the cortex.

7. **Receptive fields** of nerve cells in the dorsal root ganglion are different from those for the nerve cells in the dorsal column nuclei.
   a. An **annular receptive field**, i.e., inhibitory ring and excitatory center, is characteristic of dorsal column nerve cells and those in the thalamus.
   b. Cells with the annular receptive field are located in the primary somatosensory area (also called **Brodmann area 3**), while cells with a rectangular receptive field are located in the secondary somatosensory areas.

8. Sensory receptors respond to the basic features of a complex stimulus.
   a. The different receptors break down the stimulus into its component parts.
   b. Information about each part is transmitted along different sensory pathways to the central nervous system where it is reassembled by nerve cells in the sensory areas of the cortex.

9. Information about pain and temperature follow the pathway:

> receptor → sensory neuron → substantia gelatinosa → anterolateral pathway on opposite side → ventrobasal nuclei in the thalmus → primary somatosensory cortex

10. Inhibitory interneurons in the dorsal horn can alter the sensation of the stimulus. The interneuron can be activated by other sensory fibers or higher brain centers.

B. *The visual system detects information about the color, shape, and movement of an object.*

1. Anatomical parts of the eye include:
   a. The **retina**, which contains photoreceptors and other types of nerve cells organized into layers.
   b. **Rods** and **cones**, the photoreceptor cells, which transduce a photon of light into a more negative membrane potential that is then transformed into a series of action potentials.
   c. The **lens** and **cornea**, which bend light rays and focus them on the **fovea**.

2. The cells of the retina interact in a particular sequence:

> photoreceptor cell → **bipolar cell** → **ganglion cell**

   a. Photoreceptor cells also form synapses with **horizontal cells**, which inhibit ganglion cell activity.
   b. The ganglion cells have receptive fields shaped like an **annulus**.
   c. The spatial pattern of activated ganglion cells in the retina represents the form of the object projected onto the retina.

3. Pathways for visual information are complex.
   a. Axons from the ganglion cells synapse with neurons in the **lateral geniculate nucleus (LGN)** of the thalamus.

b. Some of the axons cross over and form connections with the LGN on the opposite side.
c. The LGN contains six distinct layers of neurons whose axons synapse with neurons in the primary visual cortex.
d. The visual field is mapped onto the retina and onto the visual cortex within which the cells are organized into columns, each responding to light impinging on either the right or the left eye.

4. The receptive field of the ganglion and LGN cells is annular, while that of the visual cortex is rectangular.
   a. The cells in the visual cortex receive inputs from cells in the LGN that have three overlapping annular receptive fields.
   b. Columns of cells in the visual cortex are sensitive to the form and orientation of the bar of light.
   c. The cells in the visual cortex synthesize information about the form of the object being viewed.

5. Color vision depends on photoreceptors called cones.
   a. Three types of photoreceptor cones (red, blue, and green) are found in the retina.
   b. Each has a visual pigment capable of absorbing a particular wavelength of light.
   c. Color perception also involves ganglion cells, which can distinguish information about different colors.
   d. Color information is transmitted to special cells in the LGN, which, in turn, convey it to cells in the **visual cortex**.

6. The fovea, composed of only cones, is the region of the retina with the greatest spatial resolution and the region where the lens tries to focus images of light.

C. *The auditory system analyzes the frequency components of sound and its location.*

1. The pathway of sound transduction in the ear is:

> outer ear → eardrum → **middle ear ossicles** (malleus, incus, stapes) → oval window → **basilar membrane** in the inner ear (cochlea)

   a. The basilar membrane is designed to respond to different frequencies of sound by changing its form.
   b. The deformations of the basilar membrane are transduced into **generator potentials** by **hair cells** that sit on the basilar membrane.
   c. The hair cells release transmitters, which generate action potentials in auditory nerve fibers that convey information about the sound to the brain.

2. The information pathway to the primary auditory cortex is:

> **auditory nerve** → cochlea nucleus in the brain → **medial geniculate nucleus** of the thalmus → **primary auditory cortex**

   a. Auditory nerve fibers respond differently to frequencies of tones. One frequency will activate a particular fiber.
   b. Neurons respond in a variety of ways to a particular tone.
   c. The fundamental features of a sound are reassembled by convergence of neurons.

3. Hearing loss may result from destruction of hair cells, damage to ossicles, brain tumors, and lesions of the auditory pathway.

D. *The gustatory system gives us an appreciation for different taste sensations.*

1. Taste receptors or gustatory receptors are located in the tongue, palate, and pharynx.

2. The receptors detect chemicals associated with four basic taste qualities: sweet, sour, salty, and bitter.

3. Each receptor responds best to only one type of taste quality but can respond to several types of substances.
   a. The binding of the chemical to the receptor produces a generator potential in the taste bud.
   b. This, in turn, causes an action potential in the afferent fiber associated with the receptor.
   c. One afferent fiber innervates many receptors.

4. Taste buds are innervated by parts of different cranial nerves.
   a. The pathway of information from receptor to the brain is:

receptor → afferent fiber → gustatory nucleus in the brain stem → ventral posterior medial nucleus of thalmus → gustatory area of neocortex

b. Alternate pathways transmit information to the limbic system and hypothalamus.

E. The olfactory system allows us to detect odor molecules in the air.

1. Olfactory receptors sensitive to different types of odors are located in the nasal cavity.

2. The binding of an odor molecule produces a generator potential, which becomes an action potential above the threshold level.

3. The pathway of information to the brain is:

receptor cell → mitral cell in olfactory bulb → olfactory tract → limbic system of cortex

Key Terms

action potential
basilar membrane
bipolar cells
cochlea
column nuclei
cones
dermatome
dorsal column pathway
dorsal horn
dorsal root ganglion
fovea
frequency code of stimulus intensity
ganglion cells
generator potential
gustatory receptor
hair cellrods
horizontal cells
lateral geniculate nucleus
limbic system
mechanoreceptors
neocortex
nocioceptors
olfactory bulb, receptor, tract
ossicles
Pacinian corpuscle
phasic discharge
photoreceptor
population code
primary somatosensory area
proprioceptors
receptive field
secondary somatosensory area
sensory adaptation
somatotopic organization
spinal cord
supporting cell
tactile receptor dorsal
taste bud
tectorial membrane
thalamus
thermal receptors
tonic discharge
tonotropic organization
ventral horn
visual agnosia
visual field

Testing Your Knowledge

I. Matching. Select the correct word from the Key Terms.

________ 1. cell bodies of sensory neurons

________ 2. part of spinal cord containing nerve cells involved in processing of sensory information

________ 3. change in membrane potential in a receptor

________ 4. part of the brain receiving all types of sensory information

________ 5. intensity of stimulus conveyed by number of action potentials per unit time

________ 6. situation when continued stimulus no longer results in perception of sensation

________ 7. detect changes in positions of limbs

________ 8. receptors detecting painful stimuli

________ 9. situation in which proprioceptor generates action potential only with change in direction of move ment

________ 10. most sensitive part of retina

________ 11. cells in retina sensitive to color

________ 12. axons of these retinal cells become the optic nerve

________ 13. area of thalamus receiving infor mation from the retina

________ 14. cell that releases transmitter in response to being moved by basilar membrane

________ 15. receptors sensitive to chemical stimulus

II. Dichotomous. Circle the correct choice from the paired terms in parentheses.

1. The Pacinian corpuscle is a (mechanorecep- tor/proprioceptive) receptor.

2. A dynamic proprioceptor generates an action potential response called a (toxic/ phasic) discharge.

3. The region of the body with the greatest density of sensory receptors is represented by the (greatest/least) proportion of the cortex.

4. Information on pain and temperature is transmitted by neurons located in the (dorsal/anterolateral) pathway.

5. Ganglion cell activity may be inhibited by (horizontal/bipolar) neurons.

6. Fibers from ganglion cells form connections with the (lateral geniculate/ventral posteri or lateral) nucleus.

7. Color and form of an object are analyzed (independently/together) in the visual cortex.

8. Cilia of hair cells are embedded in the (tectorial/basilar) membrane.

9. The intensity of a sound wave is represented by its (amplitude/frequency).

10. Conductive hearing less results from damage to the (ossicles/hair cells).

III. True/False. In the space to the left of the number write T for true or F for false.

_______ 1. The hair receptor is a rapidly adapting tactile mechanoreceptor.

_______ 2. Cell bodies of peripheral motor neurons are located in the dorsal root ganglion.

_______ 3. The image projected onto the retina is right side up.

_______ 4. Cones operate better than rods in a dark room.

_______ 5. A mirror image of the external visual field is mapped onto the visual cortex.

_______ 6. In the visual cortex, the receptive field is rectangular.

_______ 7. Bending of the stereocilia generates an action potential in the hair cell that conveys information to auditory cortex.

_______ 8. A hertz is a unit of measure for sound intensity.

_______ 9. Each gustatory fiber innervates one taste bud and responds to one stimulus.

_______ 10. The olfactory system is the only sensory system with direct innervation of the cortex.

IV. Sentence Completion. Write your response in the space provided.

1. The longitudinal orientation of the spinal cord is called the __________ axis.

2. The ____________ horns contain nerve cells involved in processing sensory information.

3. The somatic sensory cortex is part of the _________________.

4. The shape of the lens is controlled by _________________.

5. There are ______________ types of photoreceptor cones.

6. Hair cells sit on the ___________ membrane.

7. Auditory nerve fibers with a ____________ characteristic frequency innervate hair cells near the base of the cochlea.

8. The localization of sounds in space is deter mined by the ________________.

9. Information about interaural intensity and time differences is transmitted by different pathways and converges in the _________.

10. In humans the _________________ system is not as well developed as other sensory systems.

V. Multiple Choice. Circle the letter of the best response.

1. Which is not a mechanism for coding information about a stimulus?
   a. changing frequency of action potentials
   b. increasing number of receptors
   c. selective pathway
   d. increasing amplitude of action potentials
   e. all of the above

2. A term that means "close to the midline" is:
   a. dorsal
   b. medial
   c. lateral
   d. caudal
   e. ventral

3. Sensory input from the waist and thighs would be received by the:
   a. C1-C8 segments
   b. T1-T12 segments
   c. L1-L5 segments
   d. S1-S5 segments
   e. sacrococcygeal segments

4. Which sequence represents the transmission of information about a pain in the front of the right thigh?
   a. right sensory fiber → segment L2 → right anterolateral column → thalamus → cortex
   b. right sensory fiber → segment S2 → right anterolateral column → thalamus → cortex
   c. right sensory fiber → segment L2 → left anterolateral column → thalamus → cortex
   d. right sensory fiber → segment S2 → left dorsal column → thalamus → cortex
   e. right sensory fiber → segment L2 → right dorsal column → thalamus → cortex

5. Which sequence represents the neural transmission of visual information about a red ball?
   a. rod → ganglion → bipolar
   b. rod → bipolar → ganglion
   c. cone → bipolar → ganglion
   d. cone → ganglion → bipolar
   e. cone → horizontal → ganglion

6. The visual pigment in rods is:
   a. rhodopsin
   b. transducin
   c. peptide P
   d. cGMP
   e. GABA

7. In the visual cortex, cells that respond to the orientation of the receptive field make up:
   a. visual orientation columns
   b. ocular dominance columns
   c. dorsal columns
   d. anterolateral columns
   e. ventrobasal columns

8. Which sequence represents the pathway for transmission of auditory information?
   a. hair cell → auditory neuron → cochlear nucleus → mediate geniculate nucleus → primary auditory cortex
   b. hair cell → cochlear nucleus → auditory neuron → mediate geniculate nucleus → primary auditory cortex
   c. auditory neuron → hair cell → cochlear nucleus → primary auditory cortex → mediate geniculate nucleus
   d. auditory neuron → cochlear nucleus → primary auditory cortex → hair cell → mediate geniculate nucleus
   e. none of the above

9. Which sequence represents the gustatory pathway for taste buds located in the back of the tongue?
   a. cranial nerve IX → gustatory nucleus → ventral posterior medial nucleus → neocortex
   b. cranial nerve VII → gustatory nucleus → ventral posterior medial nucleus → neocortex
   c. cranial nerve X → gustatory nucleus → ventral posterior medial nucleus → neocortex
   d. cranial nerve IX → ventral posterior medial nucleus → gustatory nucleus → neocortex
   e. cranial nerve VII → ventral posterior medial nucleus → gustatory nucleus → neocortex

10. The sensation of pain is altered primarily by:
    a. interneurons
    b. ganglia
    c. hair cells
    d. nociceptors
    e. horizontal cells

VI. Complete the Table.

For each receptor, identify the stimulus and the part(s) of the spinal cord and brain involved in the transmission of information from the receptor to the neocortex.

Receptor	Stimulus	Spinal Cord or Brain Stem	Brain
rod			
cone			
hair cells in cochlea			
gustatory receptor			
olfactory receptor			
Pacinian corpuscle			
nocioceptor			

Chapter 9

Motor Systems

Chapter Learning Objectives

1. Give an overview of the interaction of nerve and muscle.
2. Indicate how the muscle spindle and Golgi tendon organs operate; describe the structure, mechanisms of action, and type of information transmitted by muscle spindles and Golgi tendon organs.
3. Describe how the spinal cord is organized.
4. Present the nature of reflex actions using the knee jerk reflex as an example.
5. Explain the inverse myotatic response.
6. Examine how the brain functions in motor responses.
7. Give the significance of the somatotopic organization of the motor and sensory cortices.
8. Indicate how sensory feedback can modify motor responses.
9. Explain the role of supplemental and promoter areas in the planning of motion.
10. Describe how posture is maintained and discuss the role of the vestibular apparatus.

Expanded Chapter Outline

I. Muscle: Controlled by Nerves

A. Motor areas of the brain transmit information about movement.

1. The motor system controls locomotion, posture, and fine movement through muscles.
   a. The motor system is organized hierarchically according to the scheme: premotor area (highest level) → **motor cortex** → brain stem → spinal cord (lowest level).
   b. The motor system also has a parallel organization in which information from two levels can affect a third level.

2. Skeletal muscle is composed of muscle fibers of two types: extrafusal and intrafusal.
   a. Skeletal muscles are usually arranged in antagonistic pairs.
   b. Skeletal muscles can be categorized as fast or slow and differ from each other in amount of vascularization, resistance to fatigue, and the functional properties of the motor neurons innervating these muscles.

*B. **Motor neurons** cause skeletal muscles to contract.*

1. Contraction of skeletal muscle cells is controlled by the alpha motor neurons in the spinal cord.

2. **Acetylcholine**, released at the neuromuscular junction, binds to **nicotinic receptors** located on the skeletal muscle cell membrane.

3. Motor neurons innervating fast and slow types of muscles differ in their frequency of discharge of action potentials and conductive velocities.

4. A **motor unit** consists of an individual motor neuron and all of the muscle fibers that it innervates. The size of the motor unit parallels the size of sensory receptive fields.

5. The amount of contractive forces generated by a muscle is controlled by the nervous system by either a frequency or population code.

II. The Spinal Cord in the Motor Response System

A. Motor neurons and muscle fibers represent functional units.

1. Motor neurons are organized in the ventral horn of the spinal cord according to the position and types of skeletal muscles innervated by these neurons.

2. Motor neurons innervating a single muscle are functionally grouped in motor neuron pools.

B. Interneurons of the spinal cord coordinate the response of antagonistic muscles.

*C. **Locomotor generators** in the spinal cord control walking movements.*

1. Interneurons and their interconnections compose the locomotor generators, which coordinate the flexor and extension of leg muscles in walking.

2. The locomotor generators are under the control of a "locomotor command center" in the brain stem.

D. Sensory fibers in skeletal muscle convey information about muscle length, force, and velocity of shortening.

1. **Sensory neurons** transmit to the central nervous system information received from the **muscle spindle** and **Golgi tendon organs**.
   a. The muscle spindle is composed of two types of intrafusal fibers that are innervated by two types of nerve fibers.
   b. Its sensitivity is maintained by gamma motor neurons, which innervate contractile elements at the poles of the spindles.

2. Golgi tendon receptors innervate tendons and transmit information about the force exerted by the muscle.

E. Spinal cord reflexes are activated by sensory receptors.

1. Spinal cord reflexes are carried out entirely within the spinal cord and are modified by input from higher centers.

2. The **stretch reflex** occurs when a muscle spindle detects a tap on the knee.
   a. The information about the stimulus is conveyed to the spinal cord.
   b. The motor neurons to the leg extensor are activated and motor neurons to the flexor are inhibited, thereby causing the leg to extend.
   c. Interneurons provide for a well-coordinated motor response.

3. The **inverse myotatic reflex** protects muscles from damage that might occur during weight lifting.
   a. The **Golgi tendon organ** detects excessive force and transmits this information to the spinal cord.
   b. There, the message is sent to inhibitory neurons, which turn off motor neurons innervating the biceps and activate motor neurons innervating the extensors.
   c. The reflex is believed to involve the smooth onset and termination of muscular contraction in walking.

4. The **flexor withdrawal reflex** involves cutaneous receptors and occurs in response to a painful stimulus.

III. The Brain in the Motor Response System

A. The brain stem controls posture.

1. Axons in the **ventromedial pathway** affect motor neurons controlling axial muscles.

2. Fibers in the **lateral reticulospinal tract** innervate flexors in the control of posture.

3. Fibers in the **rubrospinal tract** innervate motor neurons controlling distal flexor muscles.

B. The motor cortex controls reacting and fine voluntary movement.

1. The fine control of voluntary movement is due to instructions from the **motor cortex**.

2. Cells in the motor cortex are organized in a somatotopic manner.

3. The amount of cortical space devoted to different parts of the body is in proportion to the degree of fine motor control.

4. The cells of the cortex are organized in functional columns called cortical efferent zones. Neurons in the motor cortex are organized horizontally into six different layers.

5. Descending pathways from the motor cortex include:
   a. Fibers in the **corticospinal pathway** that control distal limb muscles.

b. Fibers in the **corticobulbar pathway** that control axial muscles near the midline.
c. Nerve fibers existing from the **lateral corticospinal tract** that innervate motor neuron pools controlling distal limb muscles.
d. Fibers in the **ventral corticospinal tract** that innervate motor neurons associated with axial muscles.

6. Sensory feedback to the motor cortex is by way of the somatic sensory cortex. The nerve cells in the sensory cortex are connected to those in the motor cortex in a topographic manner.

7. Nerve cells in the motor cortex encode information about direction and force required to make a reaching movement.

8. Coding of contractile force is carried out by two types of neurons: dynamic neurons and neurons coding for steady-state force.
   a. Dynamic neurons change their activity when there is an increase in the level of applied force.
   b. Neurons coding for steady-state force discharge action potentials throughout the entire period during which force is maintained.

C. The supplemental and premotor areas program movement.

1. The **supplemental and premotor cortex,** as well as the posterior parietal cortex, **cerebellum**, and **basal ganglia,** are involved in planning and programming movements.

2. The **motivational areas** of the brain, responding to the need to accomplish a particular task, transmit this information to the supplemental and premotor cortex.

D. The posterior parietal cortex integrates sensory stimuli for purposeful movement.

1. The **posterior parietal cortex** is necessary for the processing of sensory stimuli leading to purposeful movement.

2. Three types of neurons (arm projection, hand-eye coordination, and hand manipulation) in the posterior parietal cortex become active only with specific behavioral motor responses.

E. The cerebellum and the basal ganglia coordinate movements.

1. The **cerebellum and basal ganglia** both receive inputs from the neocortex and transmit information back to the cortex by way of the thalamus.

2. The cerebellum has three major functions: planning, coordination, and posture, each associated with a specific location.
   a. **Planning** of a movement is the responsibility of the lateral region of the anterior and posterior lobes.
   b. The control of **posture** and equilibrium is centered in the flocculonodular lobe.
   c. The control of limb movement involves the midline and intermediate regions of the anterior and posterior lobes.

3. **Purkinje cells**, the most prominent type of neuron, are packed tightly into a single cellular layer.
   a. Axons of Purkinje cells located in different regions of the cerebellum form synapses with neurons in one of three deep cerebellar nuclei: the dendate nucleus, the interpositus nucleus, and the fastigial nucleus.
   b. The type of information transmitted between the Purkinje cells and cells of the deep cerebellar nuclei depends on the location of the Purkinje cells.
   c. Outflow of information from the cerebellum is via the deep cerebellar nuclei but is modulated first by inhibitory actions of the Purkinje cells.

4. Inputs to the Purkinje cells occur by two routes: climbing fibers and parallel fiber pathways.
   a. **Climbing fibers** are thought to modulate the responsiveness of Purkinje cells to parallel fiber inputs.
   b. Cerebellar nerve cells can also encode the direction of reaching.
   c. Preferential direction of movement produces directional and population vectors for cells and cell clusters in the cerebellum.

5. The basal ganglia plays a role in planning movements.
   a. The caudate and putamen receive inputs

to the basal ganglia from the neocortex, thalamus, and substantia nigra of the brain stem.

b. The globus pallidus of the basal ganglia provides the output to the prefrontal and premotor cortices via the thalamus.

c. The basal ganglia are involved in the programming of motor patterns.

IV. Vestibular Control of Equilibrium and Balance

A. The vestibular system aids in maintaining the body's balance by detecting the position and motion of the head in space.

1. The **vestibular apparatus** consists of three **semicircular canals**, which detect angular acceleration in one of three planes in space, and the **utricle** and **saccule**, which detect linear acceleration.

2. When stimulated by the movement of fluid, **hair cells** in the semicircular canals and otolithic organs excite or inhibit **vestibular nerve** fibers that innervate them.

3. The information from vestibular nerves is transmitted to the vestibular nucleus, located in the brain stem; from there it is sent down to neurons regulating movement and up to the cerebellum and somatosensory cortex.

4. The cerebellum and somatosensory cortex use the information to coordinate muscular activity to maintain balance.

IV. Disorders of the Motor System

A. The type of lesion and symptoms depend on the area of the tract damaged.

B. Cerebellar lesions are characterized by a lack of coordination and equilibrium.

Key Terms

antagonistic pair
basal ganglia
brain stem
corticobulbar pathway
corticospinal pathway
extensor
extrafusal fiber
fast skeletal muscle
flexor
Golgi tendon organ
hair cell
insertion
intrafusal fiber
inverse myotatic reflex
lateral corticospinal tract
lateral reticulospinal tract
ocomotor generator
mossy fiber
motor cortex
motor neuron (alpha, gamma)
motor neuron pool
motor unit
muscle spindle
myoneural junction
origin
otolith
Purkinje cell
pyramidal tract
rubrospinal tract
saccule
semicircular canals
slow skeletal muscle
stretch receptor
stretch reflex
utricle
ventral corticospinal tract
ventromedial pathway
vestibular apparatus
vestibular nerve fiber

Testing Your Knowledge

I. Matching. Select the correct term from the Key Terms.

________ 1. neurons that directly innervate muscle cells

________ 2. strongest, most common muscle fiber

________ 3. poorly vascularized, rapidly contracting muscle

________ 4. synonymous with motor end plate

________ 5. neuron synapsing with extrafusal fiber

________ 6. functional unit of neuromuscular system

________ 7. detects length of muscle and its velocity of contraction

________ 8. detects tension in muscle

________ 9. exemplified by knee jerk

________ 10. reflex regulating muscle tension

________ 11. coordinate flexion and extension of leg muscles in walking

________ 12. location of motor neurons in spinal cord

________ 13. most prominent cell in the cerebellum

________ 14. detects angular acceleration

________ 15. detects linear acceleration

II. Dichotomous. Circle the correct choice from the paired terms in parentheses.

1. (Fast/Slow) skeletal muscles are poorly vascularized and fatigue easily.
2. A motor unit is composed of (one motor neuron and several muscle fibers/several motor neurons and one muscle fiber).
3. The number of muscle fibers in a motor unit for a thigh muscle would be (greater/less) than the number in a motor unit for a finger muscle.
4. (Excitatory/Inhibitory) effects of interneurons are associated with activation of synergistic muscles.
5. Sensory receptors that detect the length of the muscle and its velocity of contraction are called (muscle spindles/Golgi organs).
6. (Gamma/Alpha) motor neurons innervate muscle fibers.
7. The size of the motor cortex for control of the fingers is (greater/less) than the size for movement of the right leg.
8. Stimulation of the supplemental and premotor areas produces movements (more/less) complex than those produced by stimulation of the motor cortex.
9. The primary input to the basal ganglion is from the (neocortex/thalamus).
10. (Angular/Linear) acceleration is detected by the semicircular canals.

III. True/False. In the space to the left of the number write T for true or F for false.

_______ 1. The rates of action potential discharge and conduction velocity are the same for all motor neurons.

_______ 2. Epinephrine is released at a neuro muscular junction.

_______ 3. A motor unit is composed of one or more motor neuron pools.

_______ 4. The locomotor generator for the right leg is dependent on the locomotor generator for the left leg.

_______ 5. There is a direct relationship between the degree of fine motor control and amount of cortical space.

_______ 6. Motor neurons in the lateral corticospinal tract innervate motor neuron pools controlling axial muscles.

_______ 7. Purkinje cells are capable of changing their pattern of activity during adaptation of new motor responses.

_______ 8. Nerve fibers from the substantia nigra to the basal ganglia release GABA.

_______ 9. Hair cells in the vestibular apparatus generate action potentials unlike those located in the cochlea.

_______10. Hyperpolarization of a hair cell excites the vestibular nerve fibers.

IV. Sentence Completion. Write your response in the space provided.

1. The antagonistic muscle to a flexor is a(n) _______________.
2. Axons of motor neurons leave the spiral cord through the _______________ root.
3. The response of extensors and flexor muscles in the arm is coordinated by _______________.
4. Sensory feedback from the _______________ phase is necessary for the step cycle.
5. When intrafusal fibers are stretched, the discharge of action potentials in the muscle spindle _______________.
6. Axons descending from the motor cortex form the _______________ tract.
7. Sensory feedback to the motor cortex occurs by way of the _______________ cortex.
8. Somatic and visual sensory information is transmitted from the _______________ cortex to the supplemental and premotor areas.
9. _______________ cells are most prominent in the cerebellum.

10. Parkinson's disease results from a degeneration of the dopamine fibers between the _____________ and _______________.

V. Multiple Choice. Circle the letter of the best response.

1. Which of the following pairs relating location of motor neuron and type of movement is mismatched?
   a. dorsal area of ventral horn — flexor movement
   b. ventral area of ventral horn — extensor movement
   c. dorsolateral region of ventral horn — muscles in extremity
   d. ventromedial region of ventral horn — axial muscles for posture
   e. none of the above

2. Which of the following statements is not true for the muscle spindle?
   a. It detects changes in the length of the muscle.
   b. It is composed of a muscular bag and chain fibers.
   c. It acts as a stretch receptor.
   d. There is a direct relationship between stretch of muscle spindle and discharge of action potentials.
   e. It is innervated by alpha motor neurons.

3. Which of the following pairs of type of reflex and stimulus is mismatched?
   a. stretch reflex — tapping the patellar tendon
   b. inverse myotatic reflex — lifting heavy weights
   c. inverse myotatic reflex — walking
   d. flexor withdrawal reflex — stepping on a nail
   e. none of the above

4. The pathway for transmission of information about a stretched muscle is:
   a. muscle spindle → sensory neuron → dorsal root → dorsal column → somatic sensory cortex
   b. Golgi tendon → sensory neuron → ventral root → dorsal column → somatic sensory cortex
   c. muscle spindle → sensory neuron → ventral root dorsal column → somatic sensory cortex
   d. Golgi tendon → sensory neuron → dorsal root → dorsal column → somatic sensory cortex
   e. muscle spindle → sensory neuron → dorsal root → ventral column → somatic sensory cortex

5. When a person lifts an excessive weight, (the):
   a. stretch reflex is activated
   b. interneurons inhibit flexors and excite extensors
   c. interneurons excite flexors and inhibit extensors
   d. flexor withdrawal reflex is activated
   e. muscle spindle sensory neurons in the biceps inhibit the muscle from extending

6. Which of the following pathways from nerve clusters in the brain stem to motor neurons of the spinal cord is mismatched with the muscle innervated by these motor neurons?
   a. medial reticulospinal tract — activates extensor muscles for posture
   b. lateral reticulospinal tract — activates flexors for posture
   c. rubrospinal tract — activates distal flexor muscles for posture
   d. vestibulospinal tract — reflex control of equilibrium
   e. tectospinal tract — controls extensor muscles for posture

7. The major subcortical components of the motor system are the:
   a. motor cortex and cerebellum
   b. sensory cortex and cerebrum
   c. basal ganglia and cerebellum
   d. posterior parietal cortex and motor cortex
   e. brain stem and spinal cord

8. The cerebellar lobe involved in maintenance of equilibrium and posture is the:
   a. anterior
   b. posterior
   c. flocculonodular
   d. Purkinje
   e. fastigial

9. During a clockwise rotation of the head:
   a. hair cells on the left side are depolarized and those on the right side are hyperpolarized
   b. hair cells on both sides of the head are hyperpolarized to varying degrees
   c. hair cells on both sides of the head are depolarized to varying degrees

d. hair cells on the left side are hyperpolarized and those on the right side are depolarized
e. the vestibular nerves on the left side increase their rates of action potential generation and those on the right decrease

10. Information flows from the vestibular nerves directly to the:
a. vestibular nucleus
b. cerebellum
c. spinal cord
d. motor cortex
e. a and b are correct

VI. Complete the Table.

For each structure in the brain describe its location and function in the neuromuscular system.

Structure	Location	Function
supplemental cortex		
premotor cortex		
posterior partietal cortex		
anterior lobe		
posterior lobe		
flocculonodular lobe		
dendate nucleus		
interpositus nucleus		
fastigial nucleus		
inferior olivary nucleus		
mossy fibers		
caudate		
putamen		
globus pallidum		

VII. Fill in the Table.

For each tract or pathway list the origin of the axon, its location, the type of information carried, and the role in controlling movement, posture, and coordination.

Tract/Pathway	Origin	Location	Type of Information	Role
ventromedial pathway				
lateral reticulospinal tract				
rubrospinal tract				
corticospinal pathway				
corticobulbar pathway				
pyramidal tract				
lateral corticospinal tract				

Chapter 10

The Autonomic Nervous System

Chapter Learning Objectives

1. Describe the peripheral autonomic nervous system.
2. Diagram the relationship of the spinal cord and the autonomic nervous system.
3. Compare the parasympathetic and sympathetic nervous systems as to general effect, the target organs, relative lengths of preganglionic and post-ganglionic neurons, location of pre-ganglionic cell bodies, and types of neurotransmitters released by each neuron.
4. Distinguish between cholinergic and adrenergic impulses.
5. Describe the spinal autonomic reflex arc.
6. Outline the processes of micturition and defecation.
7. Explain referred pain and segmented spinal reflexes.
8. Indicate how water balance is controlled by the body.
9. Outline how breathing is regulated.
10. Explain how blood pressure is controlled.
11. Trace the fight or flight response.

Expanded Chapter Outline

I. Organization of the Autonomic Nervous System

A. *The peripheral autonomic nervous system regulates the activity of other organs in the body.*

1. The **peripheral nervous system (PNS)** is composed of the sympathetic and parasympathetic nervous system.

2. They differ in function, the types of neurotransmitters released, origin of nerves from the spinal cord, and their effects on their target organs.

B. *The central autonomic nervous system orchestrates the coordinated activity of organs to function in concert.*

C. *Neurotransmitters in the autonomic nervous system exhibit a unique pattern of distribution.*

1. Acetylcholine and norepinephrine are the two major transmitters.

2. **Cholinergic receptors** are classified as either **nicotinic** or **muscarinic**.

3. **Adrenergic receptors** are classified as alpha or beta, each of which is further subtyped into alpha1, alpha2, beta1, or beta2.
   a. The **alpha receptor** subtypes differ in their anatomical locations.
   b. The **beta receptor** subtypes differ in their responses to norepinephrine and epinephrine.

II. Control of Bodily Organs by the Sympathetic and Parasympathetic Systems

A. *The actions of the* ***sympathetic*** *and* ***parasympathetic*** *inputs to the target organ have opposite effects.*

B. *Receptors for norepinephrine and acetylcholine are found in the membranes of cholinergic and adrenergic fibers, respectively, and provide for local control at the site of the target organ.*

C. *Organs that have a vegetative function are activated by inputs from the parasympathetic system but inactivated by the sympathetic nervous system.*

D. *Organs involved in stress responses are activated by inputs from the sympathetic system but inactivated by the parasympathetic system.*

III. Reflexes Governed by the Autonomic Nervous System

A. *Spinal reflex arcs can activate autonomic functions.*

1. Information, conveyed by sensory neurons from the target organ to the spinal cord, is transmitted up to the brain and to preganglionic neurons in the **intero-medial lateral (IML) column**.

2. This loop forms the spinal autonomic reflex arc:

brain
↑
organ → sensory neuron → spinal cord →
↑ ↓
postganglionic ← interomedial lateral column
neuron

B. *Reflexes in the bladder and rectum aid in urination and defecation, respectively.*

1. The **urination reflex** involves the actions of two groups of muscles (smooth and skeletal), sensory receptors and nerves, somatic motor neurons, a higher brain center, and the ANS.

2. The **defecation reflex** involves the actions of smooth and skeletal muscles innervated by parasympathetic and somatic motor neurons, respectively, sensory stretch receptors and neurons, and higher brain centers.

C. *Pain from internal organs is localized to other parts of the body.*

1. **Referred pain** results from the convergence of nerve fibers from pain receptors in internal organs and on the surface of the body.

IV. Regulatory Systems of the Autonomic Nervous System

A. *Autonomic regulatory systems maintain a constant internal environment for optimal functioning of various organs and their cells.*

B. *Autonomic regulatory systems involve the brain stem and hypothalamus and operate by a* ***negative feedback mechanism****.*

C. *Thirst is regulated by osmoreceptors and receptors for water volume.*

1. **Osmoreceptors** in the hypothalamus and **water-volume sensors** in large veins detect changes in osmolarity and volume of the blood.

2. The hypothalamus releases a hormone, **vasopressin**, which acts on the kidneys and causes them to reabsorb more water.
   a. Information flows from peripheral sensors to centers in the brain stem, which then activate sympathetic nerves to the kidneys.
   b. The kidneys release an enzyme that activates a hormone that causes the hypothalamus to release vasopressin.

B. *Breathing is regulated by chemoreceptors and stretch receptors.*

1. **Chemoreceptors** in the ANS, called **carotid bodies** and **aortic bodies**, monitor the amounts of O_2 available to cells and CO_2 produced by cells.

2. Chemoreceptors located in the **medulla** monitor the amounts of CO_2 and H^+.

3. The peripheral and central chemoreceptors act together to coordinate the intake of oxygen and removal of CO_2, by regulating breathing according to the metabolic demands of the body.

C. *Blood pressure is regulated by baroceptors.*

1. Regulation of blood pressure involves **baroreceptors**, which detect changes in blood pressure and three brain stem centers: the **depressor center**, the **nucleus of the vagus**, and the **pressor center**.

2. The **hypothalamus** affects the regulation of blood pressure by its excitatory inputs to the pressor center and preganglionic sympathetic neurons.

V. Autonomic Functions of the Hypothalamus

A. *Activation of the* ***dorsal hypothalamus*** *enhances responses associated with feeding behaviors such as increasing intestinal mobility and blood pressure.*

B. *Activation of the* ***ventral hypothalamus*** *enhances responses associated with defense behaviors or the* ***"fight or flight responses."***

VI. Hypothalamic Control of the Adrenal Gland

A. *The cells of the **adrenal medulla** are innervated by pre-ganglionic sympathetic nerve fibers.*

1. They are homologous to post-ganglionic neurons and release **catecholamines**: norepinephrine and epinephrine.

2. The adrenal medulla coordinates the body's response to stress.
   a. In response to stress, catecholamines and glucosteroids are secreted by the adrenal medulla and adrenal cortex, respectively.
   b. The release of **glucocorticoids** depends on secretions from the hypothalamus and anterior pituitary.
   c. The release of catecholamines is carried out by direct connection of nerve fibers from the hypothalamus to the IML, from which nerve fibers enter the medulla.

Key Terms

acetylcholine
adrenal medulla
adrenergic
alpha receptor
angiotensin
aortic bodies
baroreceptors
beta receptor
carotid bodies
catecholamine
cerebrospinal fluid
chemoreceptor
cholinergic
continence
defecation reflex
depressor center
fight or flight response
hypothalamus
interomedial lateral nuclei
medulla
micturition
muscarinic receptor
nicotinic receptor
norepinephrine
osmoreceptor
pacemaker cells
post-ganglionic neuron
pre-ganglionic neuron
pressor center
sphincter
urinary bladder reflex
vasopressin (ADH)

Testing Your Knowledge

I. Matching. Select the correct term from the Key Terms.

________ 1. transmitter released from postganglionic parasympathetic neurons

________ 2. location of cell bodies of preganglionic sympathetic neurons

________ 3. receptors associated with cholinergic fibers

________ 4. receptors associated with adrenergic fibers

________ 5. ability to retain urine in the bladder

________ 6. detect changes in water volume and located in hypothalamus

________ 7. acts on kidney to increase reabsorption of water

________ 8. chemoreceptors sensitive to CO_2 and O_2

________ 9. location of central chemoreceptors sensitive to CO_2 and H^+

________ 10. detect changes in blood pressure

________ 11. regulatory center causing decrease in blood pressure

________ 12. regulatory center causing increase in blood pressure

________ 13. cells are homologous to postganglionic sympathetic neurons

________ 14. bathes brain and provides extracellular environment for brain cells

________ 15. transmitter secreted by postganglionic sympathetic fibers

II. Dichotomous. Circle the correct choice from the paired terms in parentheses.

1. Pre-ganglionic nerve cells in the sympathetic system have (long/short) axons.

2. Pre-ganglionic nerve fibers leave the spinal cord from the (dorsal/ventral) root.

3. Nerve fibers that transmit acetylcholine are called (cholinergic/adrenergic).

4. The binding of acetylcholine to muscarinic receptors on cardiac cells causes (hyperpolarization/depolarization) of these cells.

5. Binding of norepinephrine to alpha2 receptors in acetylcholine pre-synaptic terminals (inhibits/stimulates) release of acetylcholine.

6. (Angiotensin II/Antidiuretic hormone) acts directly on the kidney to retain water.

7. (Aortic bodies/Aortic arches) are chemoreceptors for CO_2 and O_2.

8. Arterial pressure <90 mm Hg would cause baroceptors to discharge (fewer/more) action potentials.

9. When blood pressure is >90 mm Hg, the (pressor/depressor) center is stimulated.

10. The adrenal (medulla/cortex) secretes glucocorticoids.

III. True/False. In the space to the left of the number write T for true or F for false.

_______ 1. In the parasympathetic system, post-ganglionic nerve cells are located in ganglion near the spinal cord.

_______ 2. Norepinephrine is released by post-ganglionic sympathetic and parasympathetic neurons.

_______ 3. Acetylcholine receptors in the post-ganglionic neuron and smooth muscle are identical.

_______ 4. Beta2 receptors may be activated by norepinephrine from post-ganglionic neurons and epinephrine.

_______ 5. The antagonistic actions of the sympathetic and parasympathetic nervous systems may be controlled at the level of release of transmitter from pre-synaptic terminals.

_______ 6. Under normal physiological conditions, the osmolarity of blood is 300 mOsm.

_______ 7. Under normal physiological conditions, the average arterial blood pressure is 90 mm Hg.

_______ 8. Stimulation of the pressor center causes an increase in heart rate.

_______ 9. Activation of the ventral hypothalamus produces responses associated with feeding behaviors.

_______ 10. The adrenal medulla secretes catecholamines.

IV. Sentence Completion. Write your response in the space provided.

1. The pre-ganglionic axons of the sympathetic nervous system exit from the __________ and __________ regions of the spinal cord.

2. __________ is released at the synaptic terminals of all neurons.

3. Binding of norepinephrine to beta receptors __________ heart rate and blood pressure.

4. The __________ center in the brain stem coordinates the emptying of the bladder.

5. __________ represent painful areas in the surface of the body that are associated with inflammations of internal organs.

6. An increase in osmolarity is detected by osmoreceptors in the __________ nucleus and __________ nucleus in the __________.

7. Renin is released by the __________ in response to __________ in water volume.

8. Stimulation of the nucleus of the vagus __________ heart rate and __________ blood pressure.

9. The action of catecholamines on the liver causes an __________ in glycogenolysis.

10. The secretion of glucocorticoids is mediated by the release of __________ from the hypothalamus.

V. Multiple Choice. Circle the letter of the best response.

1. Which of the following is true for the sympathetic nervous system but **not** true for the parasympathetic?
   a. Its nerve fibers emerge from the thoracic and lumbar regions of the spinal cord.
   b. Post-ganglionic cells have short axons and synapse with smooth muscle cells in the organ.
   c. The transmitters are acetylcholine.
   d. Pre-ganglionic nerve cell axons are long and terminate in ganglia located near the target organ.
   e. It always stimulates the target organ.

2. Under which condition would blood pressure be increased?
   a. binding of acetylcholine to muscarinic receptors on cardiac cells
   b. binding of norepinephrine to beta receptors on cardiac cells
   c. binding of acetylcholine to muscarinic receptors on smooth muscle cells of blood vessels
   d. binding of norepinephrine to beta1 receptors on smooth muscle cell of blood vessels
   e. all of the above

3. Under which condition would the urinary bladder reflex occur?
   a. activation of micturition center
   b. activation of pre-ganglionic parasympathetic nerve cells
   c. inhibition of pre-ganglionic sympathetic neurons
   d. a and b are correct
   e. a, b, and c are correct

4. Which of the following features is found in both the urinary bladder and the defecation reflexes?
   a. activation of stretch receptors in smooth muscles
   b. activation of pre-ganglionic parasympathetic neurons
   c. relaxation of internal and external sphincters
   d. involvement of higher brain center
   e. all of the above

5. Under which condition would water be retained by the kidney?
   a. presence of vasopressin
   b. osmolarity <300 mOsm in blood plasma
   c. hypo-osmolarity of blood plasma
   d. absence of renin
   e. a and d are correct

6. Which of the following pairs of sensory receptor and stimulus is mismatched?
   a. supraoptic nucleus — changes in osmolarity
   b. carotid body — changes in O_2 and CO_2
   c. aortic bodies — changes in blood flow
   d. chemoreceptor in medulla — changes in O_2
   e. baroreceptor in carotid sinus — changes in blood pressure

7. Respiration would be increased if:
   a. body temperature is decreased
   b. the partial pressure of O_2 increases
   c. blood flow increases
   d. the partial pressure of CO_2 increases
   e. the concentration of H^+ decreases

8. Under which condition would blood pressure be reduced?
   a. stimulation of nucleus of vagus
   b. stimulation of pre-ganglionic sympathetic neurons to heart
   c. stimulation of pressor center
   d. inhibition of depressor center
   e. inhibition of pre-ganglionic parasympathetic neurons to heart

9. Activation of the dorsal hypothalamus results in:
   a. increases in blood pressure
   b. increase in blood flow to skeletal muscles
   c. decrease in intestinal motility
   d. increase in intestinal blood flow
   e. "fight or flight" responses

10. Which of the following is not associated with catecholamines?
   a. increased O_2 to skeletal muscles, heart, and brain
   b. vasodilation of arterioles in skin, kidney, mesentery
   c. vasodilation of arterioles to heart
   d. increased glycogenolysis
   e. all are associated

VI. Complete the Table.

Component	General Role	Type of Neurotransmitter	Level of origin from spinal cord	Length of preganglionic neuron	Length of postganglionic neuron
Parasympathetic Nervous System					
Sympathetic Nervous System					

Chapter 11

Central Integrative Skills

Chapter Learning Objectives

1. Describe and give examples of circadian rhythms.
2. Give the details of the different phases of sleep. Describe the role of the hypothalamus in the sleep-wake cycle.
3. Explain the motivational system and how it promotes a favorable internal environment.
4. Account for weight maintainance.
5. Indicate how internal osmolarity is moni tored and maintained; describe the interrelationship between regulation of thirst, water, volume, and osmolarity.
6. Indicate the role of hormones in the development of sexual characteristics.
7. Explain the reward system.
8. Outline the differences in short-term and long-term memory.
9. Present the concepts developed by split brain experiments.
10. Outline the way the brain handles language and give some of the disorders associated with language.

Expanded Chapter Outline

I. Bodily Rhythms, Cyclic Systems, and the Hypothalamus

A. Biological clocks in the brain regulate circadian rhythms.

1. Many bodily functions and biochemical processes vary on a 24-hour schedule.

2. The **suprachiasmatic nucleus** of the **hypothalamus** is thought to be the **"master clock"** that is responsible for coordinating various biological rhythms.

B. Sleep has different stages.

1. The brain displays several different stages of electrical activity during one night of sleep.
   a. Based on changes of frequency and amplitude of the brain's electrical activity, four distinct stages of sleep have been distinguished.
   b. In stage 1 the lowest amplitude and highest frequency of electrical activity is seen, while the reverse is seen in stage4.
   c. **REM sleep**, characterized by rapid eye movements, is the period of visual dreaming.
   d. The amount of sleep and the proportion of REM and non-REM sleep vary with age.

2. The hypothalamus and brain stem are the brain areas responsible for the **sleep-wake cycle**.
   a. Electrical stimulation of the pre-optic area of the hypothalamus induces **non-REM sleep**.
   b. The suprachiasmatic nucleus receives input from the retina and provides input to the pre-optic area to entrain sleep-wake cycle.
   c. The **raphe nucleus** of the brain stem is believed responsible for REM sleep.

3. Sleep disorders include:
   a. Insomnia, a general term for inability to sleep.
   b. Narcolepsy, a sleep disorder in which an individual suddenly lapses into sleep at unexpected times, which is believed to result from activation of the REM-sleep generator.

II. Motivational Systems: Maintaining an Internal Balance and Ensuring Survival

A. Motivational systems initiate behaviors that satisfy bodily needs to ensure survival of the individual or species.

B. Motivational systems are activated by internal and/or external signals.

C. Hunger is regulated by levels of glucose and triglycerides.

1. The long-term regulation of food intake appears to involve the maintenance of body weight as reflected in the amount of adipose tissue, particularly the volume of individual fat cells.

2. Insulin plays a role in the long-term regulation of body fat and in the short-term regulation of hunger.

D. Thirst is regulated by osmolarity of the blood.

C. Sexual drive and sexual behavior can be altered by exposure to hormones during development.

1. Mating is a basic drive common to all mammals and involves the anterior hypothalamus in both sexes.
   a. Mating posture is similar among animals, although sexual behaviors are quite different.
   b. **Pheromones** activate sexual behavior in many species.

2. The development of a particular pattern of sexual behavior results from the presence or absence of androgens during critical periods of development.

3. **Androgens** are thought to exert their effect on sexual behavior through the nerve cells in the **pre-optic area of the hypothalamus**.
   a. These cells, when exposed to androgens, release **luteinizing hormone** (LH) at constant levels as adults.
   b. Non-androgenized brains develop a cyclical pattern of LH release.

III. Learning and Memory Systems

*A. **Learning** is the acquisition of knowledge, whereas **memory** is the retention of knowledge.*

B. Learning is associative and non-associative.

1. In **nonassociative learning**, adaptation to a stimulus depends on its relevance to our survival.

2. In **classical conditioning**, a type of associative learning, an animal acquires an understanding about the relationship between two stimuli.

3. In **operant conditioning**, a second type of associative learning, an animal acquires an understanding of the relationship between its own behavior and a subsequent reward (or punishment).

4. The critical factor in both types is the time interval between the appearance of the two stimuli or of a behavior and its reward.

5. The most effective types of stimuli are those relevant to survival of the animal.

C. Short-term memory is transformed into long-term storage.

1. Memory has different components and progresses through different stages.

2. Short-term and long-term memory differ in stability and length of time information is retained.

D. Different neuronal pathways are used for different types of learning and memory.

1. Learning and memory systems involve neuronal memory circuits in several areas of the brain.

2. The **hippocampus** plays a role in the trans formation of short-term to long-term memory.

3. Through **"split-brain studies"** it has been determined that different mammalian species store information in different ways in the **cerebral hemispheres**.

E. Learning causes changes in cell function.

1. The cellular mechanisms of learning involve alterations of synaptic function.

2. Habituation can be explained in terms of a decrease in synaptic transmission as a result of inactivation of Ca^{2+} channels in the presynaptic terminal.

3. The underlying mechanism of **sensitization** is called presynaptic facilitation and involves interneurons that form synapses with axons of other neurons. The facilitating interneurons enhance the release of neurotransmitter.

F. *Use and non-use of neural pathways results in maintenance or atrophy of connections.*

IV. Language Systems

A. *Comprehension and the motor aspects of language are located in the dominant hemisphere.*

1. **Language centers** are located in the **dominant hemisphere** in regions of the brain called **Wernicke's area** and **Broca's area**.
    a. Wernicke's area is responsible for comprehension of language.
    b. Broca's area is responsible for motor aspects of speech and language.

2. The pathway for information transfer of written language is:

> written word → retina → primary and secondary visual cortices → association cortex called angular gyrus → Wernicke's area (where it is recognized or comprehended) → Broca's area (where motor programs for speech reside)

B. *The non-dominant hemisphere is responsible for intonation and the emotional aspects of language.*

C. *Disorders of speech and language have anatomical origins.*

1. Aphasia may result from damage to the language centers.

2. Dyslexia may be a congenital reading disability.

V. Laterality of Brain Function

A. *Spatial abilities are lateralized to one side of the brain and appear to reside in the right hemisphere.*

B. *Male and female brains exhibit differences of laterality.*

Key Terms

androgen
aphasia
associative learning
Broca's area
circadian rhythm
conditioning (operant, classical)
dyslexia
electroencephalogram (EEG)
habituation
hypothalamus
insomnia
learning
luteinizing hormone
memory (long-term, short-term)
motivational system
narcolepsy
non-associative learning
non-REM sleep
pheromone
pre-optic area
pre-synaptic facilitation
raphe nucleus
REM sleep
sensitization
split-brain studies
substantia nigra
suprachiasmatic nucleus
supraoptic nucleus
Wernicke's area

Testing Your Knowledge

I. Matching. Select the correct term from the Key Terms.

________ 1. coordinates various biological rhythms

________ 2. measures brain activity

________ 3. stage of sleep during which visual dreaming occurs

________ 4. electrical stimulation of this area induces non-REM sleep

________ 5. believed responsible for generation of REM sleep

________ 6. one of brain centers for sexual drive

________ 7. location of osmoreceptors

________ 8. hormone affecting development of sexual behavior

________ 9. area of brain associated with reward system

________ 10. inability to sleep

________ 11. decrease of response to repeated stimulus

________ 12. learning resulting in understanding of relationship between two stimulus

________ 13. area responsible for comprehension of language

________ 14. area responsible for motor aspects of speech and language

_________ 15. language disorder resulting from damage to brain

II. Dichotomous. Circle the best choice among the paired terms in parentheses.

1. During the onset and programming of sleep, the frequency of electrical activity (increases/decreases) as the amplitude (increases/decreases).

2. The functions associated with stage 4 sleep reflect activity of the (parasympathetic/ sympathetic) nervous system.

3. Short-term regulation of food intake involves (glucose/triglycerides) as a stimulus.

4. Osmoreceptors are located in the (suprachiasmatic/supraoptic) nucleus.

5. The (anterior/posterior) hypothalamus houses a brain center for sexual drive.

6. Exposure to (androgens/estrogens) during critical periods triggers male development and sexual behavior.

7. The acquisition of knowledge is called (learning/memory).

8. Recalling an address of an old friend is an example of (short-term/long-term) memory.

9. The (hypothalamus/hippocampus) plays a role in the transformation of short-term to long-term memory.

10. In a majority of left-handed persons, Wernicke's area and Broca's area are located in the (left/right) hemisphere.

III. True/False. In the space to the left of the number write T for true or F for false.

_______ 1. Body temperature is lowest at midday.

_______ 2. Changes in body function occurring in REM sleep reflect activation of the sympathetic division of the ANS.

_______ 3. The amount and proportion of REM and non-REM sleep remain constant throughout a person's life.

_______ 4. Narcolepsy is associated with a defect in the raphe nucleus.

_______ 5. Lordosis is an indication of male sexual behavior.

_______ 6. Nerve cells in the preoptic area exposed to androgen release luteinizing hormone in a cyclic pattern.

_______ 7. A decreased response to a repeated stimulus is called sensitization.

_______ 8. In the conditioning process of the eye-blink reflex by a sound, Purkinje cell activity eventually decreases.

_______ 9. The hippocampus is involved in all memory circuits.

_______ 10. In a left-handed person, the right hemisphere is always dominant.

IV. Sentence Completion. Write your response in the space provided.

1. Cycles that vary daily are called _________.

2. The "master clock" for coordinating various biological rhythms is thought to be the _________________ of the hypothalamus.

3. Control of hunger is centered in the ___________________.

4. A blood plasma osmolarity of 400 would stimulate the ___________ and ___________ to release _________________.

5. ________________ are stimuli for activating sexual behavior in many species.

6. ___________ reflects an increase in response to a stimulus.

7. The process of forming long-term memory from short-term is ___________________.

8. In _______________ conditioning, an animal learns the relationship between two stimuli.

9. In __________ conditioning, an animal gains an understanding of the relationship between its behavior and subsequent reward.

10. _________ is responsible for comprehension

of language and __________ is responsible for motor aspects of speech and language.

V. Multiple Choice. Circle the letter of the best response.

1. Stage 4 sleep is characterized by:
   a. lowest amplitude, lowest frequency of electrical activity
   b. lowest amplitude, highest frequency of electrical activity
   c. highest amplitude, lowest frequency of electrical activity
   d. highest amplitude, highest frequency of electrical activity
   e. increase heart rate and blood pressure

2. Sleep may be induced by injection of into the pre-optic area.
   a. serotonin
   b. amphetamine
   c. GABA
   d. dopamine
   e. norepinephrine

3. Which of the following plays a role in long-term regulation of food intake?
   a. amount of adipose tissue
   b. insulin
   c. ventral hypothalamus
   d. a and b
   e. a, b, and c

4. A female mouse located between two males in the uterus would, as an adult:
   a. have a cyclic estrous
   b. have constant production of LH
   c. exhibit lordosis
   d. have no estrous cycle
   e. none of the above

5. Consider the following: a chimpanzee presses a level that releases a banana. This is an example of:
   a. sensitization
   b. habituation
   c. operant conditioning
   d. classical conditioning
   e. non-associative learning

6. In the conditioning of eye-blink reflex by a sound, the critical memory circuits are connected in the:
   a. cerebellum
   b. suprachiasmatic nucleus
   c. pre-optic nucleus
   d. medulla
   e. cerebrum

7. Which of the following pairs of behavioral response and cellular mechanism is mismatched?
   a. habituation — decrease in synaptic transmission due to increase of Ca^{2+} influx
   b. sensitization — increase in synaptic transmission due to influx of Ca^{2+}
   c. classical conditioning — increase in synaptic transmission due to influx of Ca^{2+}
   d. a and b are correct
   e. a, b, and c are correct

8. Which of the following statements about topographical maps is **not** true?
   a. The maps never change.
   b. There is a direct correlation between the use of a finger and the amount of the cortex representing it.
   c. If a finger is severed, the cortical space devoted to that finger is gradually eliminated.
   d. a and b are true
   e. a, b, and c are true

9. The area of the brain believed responsible for the association of visual, auditory, and tactile sensations is (the):
   a. Wernicke's area
   b. Broca's area
   c. angular gyrus
   d. planum temporale
   e. substantia nigra

10. Damage to the region in the non-dominant hemisphere comparable to Wernicke's area results in:
    a. an inability to understand intonation
    b. an inability to express intonation
    c. an inability to speak fluently and grammatically
    d. an inability to understand spoken and written language
    e. dyslexia

VI. Complete the Table.

List the function and location for each structure or region.

Structure/Region	Function	Location
suprachiasmatic nucleus		
preoptic area		
raphe nucleus		
supraoptic nucleus		
anterior hypothalamus		
substantia nigra		
hipocampus		
Wernicke's area		
Broca's area		
angular gyrus		
corpus callosum		

Chapter 12

Endocrine Control Mechanisms

Chapter Learning Objectives

1. Indicate the relationship between the nervous system and the endocrine system; compare the nervous and endocrine systems as to speed and nature of the response, mechanism of affecting the response, and the role of receptors.
2. Describe positive and negative feedback control over hormone secretion.
3. List the different hormones of the body, the glands that produce them, the target organs, and the biological effects of the hormones.
4. Characterize the different kinds of hormones, and indicate how their chemical nature affects their function.
5. Summarize the production of steroid hormones.
6. Outline how hormones are transported by the blood. Describe factors affecting the dynamic equilibrium between free and bound hormones.
7. Explain how receptors on the plasma membrane function.
8. Detail how the second messenger system works.
9. Compare and contrast the differences in hormones that act on membrane receptors and those that enter a target cell.
10. Compare peptide and steroid hormones as to their mechanism of action, particularly regarding the nature and location of receptors and involvement of and types of second messengers.
11. Evaluate the amplification phenomenon.

Expanded Chapter Outline

I. General Concepts of Endocrine Control

A. *Hormones are chemical signals used in cell-to-cell communication.*

1. Hormones are generally considered to be substances usually secreted in small amounts into the blood by specialized cells or glands.
   a. Hormones are carried by blood to other parts of the body.
   b. They bind to specific receptors on cells in target tissues to produce a biological response.
2. Some hormones diffuse to their site of action, which may be the cell producing the hormone **(autocrine mechanism)** or to adjacent cells **(paracrine mechanism)**.

B. *The nervous system and the endocrine system function together to promote homeostasis.*

1. **Neuroendocrine cells** function as an interface between the nervous and endocrine systems.
2. They convert electrical signals from nerve cells into hormonal signals.

C. *Feedback regulation is an important feature of endocrine communications.*

1. Hormone production by endocrine cells is regulated, in most instances, by a **negative feedback mechanism**.
2. The rate of hormone secretion is regulated to maintain **homeostasis**.

D. *Hormones regulate or influence a variety of processes in the body:*

1. Energy production, use, and storage.
2. Salt and water metabolism.
3. Growth and development.
4. Reproductive function.

II. The Chemistry of Hormones

A. Most hormones can be characterized as belonging to one of three chemical classifications.

1. Hormones derived from amino acids are known as **amines** and include **epinephrine** and **melatonin**, which are derived from tyrosine and tryptophan, respectively.
   a. Amine production and secretion are triggered by the same stimulus.
   b. Thyroxine, composed of four iodide molecules joined to a pair of tyrosines, is stored as a larger precursor protein, known as thyroglobulin.

2. Most hormones are either peptides (e.g., TRH, ACTH, PTH) or proteins (e.g., insulin).
   a. Some contain a significant amount of carbohydrate (e.g., FSH, TSH, LH, HCG).
   b. Many peptide hormones are first synthesized as larger precursor molecules.

3. **Cholesterol** is the precursor for all **steroid hormones** and vitamin D.

4. Steroid hormones are grouped into five categories according to the type of biological process they regulate:
   a. **Glucocorticoids**.
   b. **Mineralocorticoids**.
   c. **Androgens**.
   d. **Estrogens** and **progestins**.

5. There are several common steps in the formation of each of the five classes of steroids.

6. The type of steroid synthesized depends on the relative amounts of the specific enzymes needed to produce the particular type of steroid.

7. Genetic defects leading to deficiencies in single enzymes of steroid synthetic pathways can cause endocrine abnormalities.

8. In contrast to the situation with other hormones, steroids are synthesized and secreted only on demand, and their production depends on the availability of cholesterol.

B. Some hormones are carried in the blood via transport proteins.

1. Mechanisms are needed to transport hormones because of their limited solubility in aqueous solutions or the possibility of urinary loss owing to small size.

2. Thyroxine and steroids bind to two types of **transport proteins**:
   a. Those with high specificity and high affinity for one hormone.
   b. Those without high specificity or affinity, such as **albumin** and **prealbumin**.

3. Albumin and prealbumin are synthesized in the liver and play an important role in the transport of some steroid hormones and thyroxine.

4. There is a dynamic equilibrium between free and bound hormones.
   a. The free hormone is the physiologically active form and is able to move from the blood to target cell receptors.
   b. **Carrier proteins** release hormones to maintain a relatively constant concentration of the free hormone in the blood.

C. Peripheral transformation, degradation, and excretion of hormones are important considerations regarding hormone action.

1. For most hormones, the form that is secreted is the active one.

2. The kidneys and liver are the major pathways for removal or degradation for most hormones.

D. Measurement of hormone concentrations in the blood provides important information to the clinician and the researcher.

III. Mechanisms of Hormone Action

A. Hormones produce their effects by one of two mechanisms of action involving either intracellular or membrane receptors.

1. **Receptors** are a key element in the initiation of hormone action.
   a. Hormones and their receptors have complementary shapes.
   b. Binding between receptor and hormone involves weak chemical forces and is usually readily reversible.
   c. For any particular hormone there is a

finite number of receptors in its target tissue and a maximal response produced.

2. The receptor serves as a **signal transducer** converting a specific extracellular hormonal signal into an intracellular signal that will redirect or alter cell function.

B. *The **"second messenger" model of hormone action** describes how hydrophilic hormones produce their effect.*

1. Peptide hormones use intracellular signals or second messengers to produce their effects. The binding of peptide hormone to its membrane-bound receptor results in the generation of the second messenger.

2. Three intracellular second messenger systems have been described:
   a. The adenylate cyclase/cyclic AMP system.
   b. The phosphatidyl inositol and diacylglycerol (DG) system.
   c. Receptor-linked ion channels

3. The activity of **adenylate cyclase** is coupled to membrane receptors by **G-proteins**, integral membrane proteins.
   a. Binding of hormone to $\mathbf{G_S}$ **protein** stimulates adenylate cyclase and increases intracellular **cAMP**, while binding of hormone to $\mathbf{G_i}$ **protein** inhibits adenylate cyclase and results in a reduction of cAMP.
   b. ACTH, epinephrine, glucagon, LH, PTH, TSH, and FSH use cAMP.

4. Binding of hormone to another type of G protein stimulates **phospholipase C** resulting in the hydrolysis of $\mathbf{PIP_2}$ to $\mathbf{IP_3}$ and diacylglycerol, which function as second messengers.
   a. IP_3 causes an increase in the intracellular concentration of calcium, which plays an important role as a mediator in a variety of different processes.
   b. DG activates, **C-kinase**, a membrane-bound protein, which catalyzes the phosphorylation of proteins that produce specific changes in cell function.

5. Binding of hormone to its receptor results in the opening of specific, gated, ion channels, which generates a signal in one of two ways depending on cell type.
   a. A transient flux of ions alters membrane potential in electrically active cells.
   b. A major influx of ions initiates a cellular response in cells that are not electrically active.

C. *The "gene expression" model of hormone action describes how hydrophobic hormones produce their effect.*

1. Steroids and thyroxine bind to **intracellular receptors** in contrast to peptide hormones.

2. Receptors for steroids and thyroid hormones are located in the nuclear envelope within the nucleus.

3. The **hormone-receptor** complex interacts with specific "acceptor sites" in nuclear DNA, thus affecting differential transcription and the production of specific hormones.

4. Responses to steroid or thyroid hormones last longer than responses to peptide hormones.

5. The time of action for hormones using the "second messenger" model and those using the "gene expression" model differ consistently.

6. Signal amplification of hormone action can occur at various steps, such as binding of the hormone to its receptor to the final biological effect produced by the hormone.

7. Amplification of hormone signal is an important aspect of hormone action.

Key Terms

adenylate cyclase
albumin, prealbumin
amine
autocrine
cholesterol
cyclic AMP
diacylglycerol
dynamic equilibrium
feedback (negative, positive)
G-protein
glucocorticoid
inositol triphosphate (IP_3)
insulin, glucagon
mineralocorticoid
paracrine
peptide hormone
phosphatidylinositol 4,5-bisphosphate (PIP_2)
phospholipase C
protein kinases
radioimmunoassay
receptor-mediated endocytosis
second messenger
signal transducer
steroid
thyroxine

Testing Your Knowledge

I. Matching. Select the correct term from the Key Terms.

_________ 1. lipid hormone, composed of four interlocking rings

_________ 2. transports steroid and thyroxine in plasma

_________ 3. hormones such as epinephrine that have a free amino group

_________ 4. contains iodide and derived from tyrosine

_________ 5. describes hormones composed of amino acids

_________ 6. describes relationship between free and bound hormones

_________ 7. internalization of hormone-receptor complex

_________ 8. second messenger whose production depends on G_S and G_i.

_________ 9. activation of phospholipase C results in production of these second messengers

_________ 10. second messenger, which increases intracellular concentration of calcium ions

_________ 11. hormone with intracellular receptor

_________ 12. hormone regulating salt and water content

_________ 13. hormone regulating concentration of plasma glucose

_________ 14. chemical molecules from which steroids derived

_________ 15. describes functional aspect of receptor

II. Dichotomous. Circle the correct choice from the paired terms in parentheses.

1. (Autocrine/Paracrine) regulation involves the secretion and diffusion of a hormone that acts in adjacent cells.
2. Radioactive iodine injected into the body would eventually be found in (thyroxine/thyrotropin).
3. Aldosterone and cortisol are produced in the adrenal (medulla/cortex), but epinephrine is produced in the adrenal (medulla/cortex).
4. The (free/bound) form of thyroxine is physiologically active.
5. Hormones are present in the blood in very (high/low) concentrations.
6. Testosterone and aldosterone would use (cell surface/intracellular) receptors.
7. The binding of a hormone to its receptor is (reversible/irreversible).
8. cAMP would be increased by binding of a hormone to the (R_i/R_s) receptor.
9. The intracellular concentration of Ca^{2+} is (raised/lowered) by the action of IP_3.
10. Responses to steroid and thyroid hormones occur (more/less) rapidly than responses to hormones using the second messenger system.

III. True/False. In the space to the left of the number write T for true or F for false.

_______ 1. Neuroendocrine cells respond to electrochemical signals from the endocrine system.

_______ 2. Each hormone has one specific target cell and biological effect.

_______ 3. Epinephrine and melatonin are synthesized by the same mechanisms as thyroxine.

_______ 4. For peptide hormones there is usually a direct relationship between secretion of the hormone and an increased rate of synthesis.

_______ 5. Steroid hormones are synthesized, stored in vesicles, and released on stimulation.

_______ 6. Each receptor forms a complex with one specific hormone.

_______ 7. Peptide hormones are hydrophobic and bind to intracellular receptors.

_______ 8. Hormone membrane receptors show lateral mobility.

_______ 9. G_s protein inhibits adenylate cyclase and leads to a decrease in intracellular cAMP.

_______ 10. The activity of pre-existing proteins is modified by hormones using the second messenger system.

IV. Sentence Completion. Write your response in the space provided.

1. Epinephrine belongs to a subset of hormones known as _______________.
2. Thyroxine is synthesized from the precursor __________________.
3. All steroid hormones are derivations of __________________.
4. The rate-limiting step in steroid synthesis is the conversion of cholesterol to __________.
5. Carrier proteins are synthesized in the __________________.
6. For most hormones the major pathways for removal and degradation occur in the ____________ and ____________.
7. The radioimmunoassay is based on the principle of __________________.
8. Insulin and epinephrine bind to _________ receptors, which generate second messengers.
9. Receptors act as __________ __________ to convert a hormonal signal to an intracellular signal.
10. Although present in low concentrations, hormones are effective regulators because ___________ occurs at nearly every step in the hormone action.

V. Multiple Choice. Circle the letter of the best response.

1. All of the following are characteristics of the endocrine system except:
   a. multilevel feedback regulation
   b. chemical messengers
   c. binding to receptors
   d. synaptic connections between cells
   e. secretion into and transport by blood

2. Hormones generally may be characterized as:
   a. amino acid derivatives
   b. peptides or proteins
   c. steroids
   d. nucleic acids
   e. a, b, and c are correct

3. Which of the following pairs of hormone: chemical class/derivative is mismatched?
   a. Thyroxine — amino acid
   b. epinephrine — steroid
   c. insulin — protein
   d. estradiol — steroid
   e. growth hormone — peptide

4. Which of the following triplets of hormone-source-target cell is mismatched?
   a. prolactin — posterior pituitary — breast
   b. glucagon — pancreas — liver
   c. aldosterone — adrenal cortex — kidney
   d. estrogen — ovary — ovary
   e. calcitonin — thyroid — bone

5. Which of the following is true for FSH, TSH, LH, and hCG and **not** other hormones?
   a. They are peptides and proteins.
   b. They are synthesized on rough ER.
   c. They bond to external cellular receptors.
   d. They are released by exocytosis.
   e. They are glycoproteins.

6. A person has a genetic defect and lacks the enzyme necessary for production of progesterone. Which of the following hormones would not be produced?
   a. aldosterone
   b. cortisol
   c. estrogen
   d. a and c
   e. a, b, and c

7. Which of the following is not true for hormones and carrier proteins?
   a. All hormones are transported in the blood via carrier proteins.
   b. Albumin binds hormones without high specificity or affinity.
   c. Thyroxine may be carried by TBG or transthyretin.
   d. The bound form of the hormone is

physiologically active.
e. a and d

8. Which of the following pairs of hormone and model of mechanism of action is mismatched?
a. glucagon — second messenger
b. aldosterone — gene expression
c. epinephrine — second messenger
d. estradiol — second messenger
e. cortisol — gene expression

9. Hydrolysis of PIP_2 to IP_3 and DG is mediated by:
a. phosphodiesterase
b. adenylate cyclase
c. phospholipase C
d. protein kinase
e. Ca^{2+}

10. Which of the following is characteristic of the second messenger system and not the intracellular receptor system?
a. Transcriptional activity of DNA is stimulated.
b. New proteins are synthesized.
c. Cellular response occurs more rapidly.
d. Long-term effects continue after removal of hormonal stimulus.
e. The cellular response to hormonal stimulation is delayed.

VI. Complete the Table

Hormone	Gland or Source	Chemical Nature	Type of Receptor	Type of Second messenger	Target Tissue	Primary Biological effect
growth hormone						
insulin						
glucagon						
aldosterone						
thyroid hormones						
androgens						
parathyroid hormone						

Chapter 13

The Pituitary Gland

Chapter Learning Objectives

1. Describe the gross and functional anatomy of the pituitary gland, and indicate how the embryonic development specifically influenced this gland.
2. Explain the close connection of the posterior pituitary with the hypothalamus and nervous tissue.
3. List the hormones produced by the anterior pituitary and give a brief description of the role of each hormone. List the hormones secreted by the posterior pituitary and their primary effect on target cells.
4. Describe how the secretion of the hormones is controlled.
5. Describe the chemical nature of the hormones associated with the pituitary gland.
6. Indicate the relationship between somatostatin and somatomedins.
7. Describe the abnormalities associated with excess or insufficient growth hormones.
8. Detail the gross anatomy of the thyroid gland and the way that the follicles are organized to lead to the production of thyroid hormones.
9. List the effects of thyroid-stimulating hormone on thyroid hormone production and the effects of the thyroid on different parts of the body. Diagram the hypothalamic-pituitary-thyroid axis, using arrows to indicate negative feedback of thyroid hormone production.
10. Indicate the problems associated with hypothyroidism and hyperthyroidism.

Expanded Chapter Outline

I. Anatomy of the Pituitary and Its Relation to the Hypothalamus

A. *During development, the pituitary is formed from two different tissue sources.*

1. The pituitary gland is composed of three lobes:
   a. **Anterior lobe** (a true endocrine gland)
   b. **Pituitary lobe** (an extension of the nervous system)
   c. **Intermediate lobe** (little or no endocrine function in humans)

2. The pituitary is connected to the hypothalamus by a stalk.

3. Both vascular and neural components are important for the regulation of pituitary functions.
   a. The posterior pituitary contains axons and terminals of neurons whose cell bodies are located in the hypothalamus.
   b. The anterior pituitary receives messages from the hypothalamus via the **hypothalamic-pituitary portal system** in the median eminence.
   c. The short portal system connects the anterior and posterior lobes.

II. Regulation of Pituitary Hormone Secretion by the Hypothalamus

A. *Hormone secretion by the posterior pituitary occurs by a typical endocrine regulatory mechanism.*

1. The posterior pituitary secretes **oxytocin** and **antidiuretic hormone** (ADH or vasopressin), which are synthesized in the cell bodies of the neurons in the hypothalamus.

2. Synthesis of these hormones depends on neural inputs to the hypothalamus from other areas of the hypothalamus and the brain.

B. *Hormone secretion by the anterior pituitary occurs by a typical endocrine regulatory mechanism.*

1. The anterior pituitary synthesizes six different hormones whose secretion is under the control of different releasing/inhibiting

hormones produced in the hypothalamus.

2. The **hypothalamic inhibitory/releasing factors** travel via the hypothalamic-pituitary portal systems to the anterior pituitary.

III. Actions and Effects of the Pituitary Hormones

A. *Hormones (aldosterone, oxytocin) are secreted from the posterior pituitary.*

1. The primary effect of aldosterone (ADH), a small peptide, is to increase water retention by the kidney.

2. ADH secretion is affected primarily by changes in:
   a. Osmolarity of the blood or extracellular fluid.
   b. Blood volume.
   c. Other factors, such as certain drugs and alcohol.

3. **Oxytocin** stimulates the release of milk from the alveoli, the sites of milk production, into the larger ducts of the breast, as well as the contraction of the smooth muscle during labor.

B. *Five cell types are responsible for the secretion of six hormones from the anterior pituitary.*

1. Within the anterior pituitary, there are five different hormone-secreting cell types:
   a. **Somatotrophs,** which produce growth hormone.
   b. **Corticotrophs,** which produce adreno-corticotropic hormone (ACTH).
   c. **Lactotrophs,** which produce prolactin.
   d. **Thyrotrophs,** which produce thyroid-stimulating hormone (TSH).
   e. **Gonadotrophs,** which produce both follicle-stimulating hormone (FSH) and luteinizing hormone (LH).

2. The size and activity of these cells vary under different physiological conditions.

3. The primary effects of **LH** and **FSH,** the gonadotropins, are:
   a. To promote maturation and development of the sperm and egg.
   b. To stimulate production of sex steroid hormones by the gonads.

4. Secretion of **prolactin** is under dual control of:
   a. **Prolactin-releasing hormone (PRH).**
   b. **Prolactin-inhibiting hormone (PIH).**

5. **ACTH** regulates the synthesis and secretion of **cortisol** from the adrenal cortex. The secretion of ACTH is regulated by **cortico-tropin-releasing hormone (CRH),** produced in the hypothalamus.

6. Growth hormone is structurally similar to **prolactin** and **human chorionic soma-tomam-motropin** secreted by the placenta.
   a. Secretion of growth hormone is regulated by **growth hormone–releasing hormone** and **somatostatin,** both of which are produced in the hypothalamus.
   b. Several factors or physiological conditions affect secretion of growth hormone through their effects on the brain and hypothalamus.
   c. Some of the effects of growth hormone are similar to those of **insulin** (insulin-like effects), whereas some antagonize or impair the actions of insulin (anti-insulin effects).

7. **Insulin-like effects** include:
   a. Uptake of amino acids and stimulation of protein synthesis in muscle cells.
   b. Stimulation of protein synthesis in the liver.

8. **Anti-insulin effects** include:
   a. Decrease of glucose into adipose tissue and muscle.
   b. Increase of gluconeogenesis in the liver.

9. Growth hormone (GH) stimulates the production of **somatomedins** produced predominantly by the liver. The two most abundant are:
   a. **Insulin-like growth factor I (IGF I).**
   b. **Insulin-like growth factor II (IGF II).**
   c. The production of somatomedins is affected by the concentration of growth hormone, insulin, and thyroid hormones.

10. IGFs have two types of effects:
   a. Stimulation of events leading to an increase in linear growth of the skeleton.
   b. Stimulation of processes, such as protein and RNA syntheses, needed for tissue growth and repair.
   c. The particular clinical symptoms associated

with deficiency or excess of growth hormone depend on the age of the individual when the deficiency or excess occurs.

11. The **thyroid gland** consists of cells arranged in closely packed follicles.
    a. Large amounts of **thyroglobulin**, the thyroid hormone precursor, are stored in the follicles to ensure sustained thyroid hormone secretion in the absence of dietary iodine.
    b. Iodine is trapped in follicle cells, where it undergoes oxidation and joins with tyrosine residues to form iodinatedtyrosines within the thyroglobulin protein.
    c. Coupling of DIT and MIT residues in different combinations produces T_3 and T_4, which are released when the thyroglobulin molecules are enzymatically degraded.
    d. **TSH** is the primary regulator of thyroid follicle cell activity and thyroid hormone secretion.
    e. The relationship between the hypothal amus, anterior pituitary, and thyroid gland is known as the **hypothalamic-pituitary-thyroid axis.**
    f. The secretion of TSH is regulated primarily by **thyrotropin-releasing hormone (TRH)**, whose secretion from the hypothalamus can be influenced by a variety of factors.
    g. Thyroid hormones exert negative feedback effects on the thyroid gland, thus limiting their own production.
    h. Thyroid hormones are transported in the plasma by carrier proteins such as thyroxine-binding globulin and albumin.
    i. T_3 is more active than T_4 within the target cell.
    j. Thyroid hormones affect many tissues and metabolic processes. These are listed in Table 13-6 in the textbook chapter.

Key Terms

adrenocorticotropic hormone
anterior pituitary
antidiuretic hormone
corticotropin-releasing hormone
cretinism
dwarfism
follicle stimulating hormone
gigantism
gonadotropin-releasing hormone
growth hormone
growth hormone-releasing hormone
hyperthyroidism
hypothalamic-pituitary portal system
hypothalamus
hypothyroidism
insulin-like growth factors (IGF-I,IGF-II)
luteinizing hormone
myxedema
oxytocin
posterior pituitary
prolactin
prolactin inhibitory hormone
prolactin releasing hormone
somatomammotropic hormone
somatomedin
somatostatin
supraoptic paraventricular nuclei system
thyroglobulin
thyroid-stimulating hormone
thyrotropin-releasing hormone
thyroxine
thyroxine binding globulin

Testing Your Knowledge

I. Matching. Select the correct term from the Key Terms.

________ 1. forms from Rathke's pouch, produces growth hormone

________ 2. forms as outpocketing from hypothalamus

________ 3. location of cell bodies of neurons secreting oxytocin

________ 4. composed of two capillary networks and through which TRH moves

________ 5. inhibits release of growth hormone

________ 6. secretion is increased when osmolarity increases

________ 7. stimulates contraction of smooth muscle in breast and uterus

________ 8. destruction of gonadotrophs would result in deficiency of which hormone(s)

________ 9. inhibited by REM sleep and high blood glucose concentration

________ 10. stimulated by low blood glucose concentration and deep sleep

________ 11. production stimulated by growth hormone

_________ 12. stimulates skeletal growth by increasing formation of cartilage

_________ 13. results from deficiency of growth hormone or somatomedin

_________ 14. precursor of thyroid hormones

_________ 15. deficiency results in no production of T_3

_________ 16. results from hypothyroidism

II. Dichotomous. Circle the best choice among the paired terms in parentheses.

1. The (anterior/posterior) pituitary develops from Rathke's pouch.
2. Oxytocin is secreted from the (anterior/posterior) lobe of the pituitary and is transported via (axons/capillaries).
3. A(n) (decrease/increase) in osmolarity stimulates release of vasopressin.
4. Oxytocin (increases/decreases) the availability of milk to a nursing infant.
5. A nursing infant would stimulate the release of (PRF/PIH) from the hypothalamus.
6. Growth hormone is considered a(n) (catabolic/anabolic) hormone.
7. (Somatomedin/Somatostatin) inhibits the secretion of growth hormone.
8. An individual without the anterior lobe of the pituitary would have a(n) (increased/decreased) production of IGF I.
9. TSH receptors are located on the (basal/apical) surface of follicle cells.
10. (TRH/TSH) directly stimulates the secretion of thyroid hormone.

III. True/False. In the space to the left of the number write T for true or F for false.

_______ 1. The anterior and posterior pituitary lobes develop from the same embryonic tissue.

_______ 2. The secretion of oxytocin and vasopressin is regulated directly by the neural input to the supraoptic and paraventricular nuclei.

_______ 3. Releasing/inhibiting hormones are synthesized by cell bodies scattered throughout the hypothalamus.

_______ 4. ADH may act as a vasoconstrictor.

_______ 5. Oxytocin is synthesized by the anterior lobe.

_______ 6. Somatostatin stimulates the release of growth hormone.

_______ 7. IGF II is primarily responsible for stimulating tissue growth and repair.

_______ 8. Iodination of tyrosine occurs in the colloid.

_______ 9. When the level of thyroid hormone is increased, the level of TSH is as well.

_______ 10. T_4 is more active than T_3 in the target cell.

IV. Sentence Completion. Write your response in the space provided.

1. The ___________ pituitary lobe is regulated by hormones transported in a complex of blood vessels known as the ______________ portal system.
2. A decrease in blood volume would stimulate release of __________________.
3. Oxytocin acts primarily in the ___________ and _______________.
4. The hormones ___________ and _________ are necessary for production and release of milk.
5. Growth hormone is produced by _________ cells in the anterior lobe.
6. The secretion of ACTH is regulated by ______________ from the hypothalamus.
7. IGF-I promotes bone growth by stimulating _____________, located in the ___________ _____________.

8. Iodine is complexed to ________________ residues in thyroglobulin.

9. Thyroid hormone is regulated by a _______ ________ feedback mechanism.

10. In the target cell, T ________________ is converted to T ________________.

V. Multiple Choice. Circle the letter of the best response.

1. All of the hormones are secreted by the anterior lobe of the pituitary except:
   a. oxytocin
   b. thyroid-stimulating hormone
   c. prolactin
   d. growth hormone
   e. luteinizing hormone

2. Which of the following pairs of hypothalmic-releasing hormone and effect on pituitary secretion is mismatched?
   a. CRH — stimulates ACTH secretion
   b. GHRH — inhibits GH secretion
   c. TRH — stimulates TSH secretion
   d. PIH — inhibits prolactin secretion
   e. GnRH — stimulates secretion of FSH

3. Under which conditions would ADH release be stimulated?
   a. increase in blood volume
   b. hemorrhage
   c. alcohol intoxication
   d. decrease in osmolarity of body fluids
   e. all of the above

4. Which of the following is **not** true for ACTH?
   a. It is derived from POMC peptide.
   b. It is produced in the paraventricular nucleus of the anterior lobe.
   c. It stimulates the release of cortisol.
   d. It acts in the adrenal cortex.
   e. All are true.

5. Under which condition would the secretion of GH increase?
   a. deep sleep
   b. hypoglycemia
   c. emotional stress
   d. presence of GHRH
   e. all of the above

6. All of the following are effects of growth hormone except:
   a. uptake of amino acids
   b. synthesis of liver proteins
   c. increased breakdown of fat stores
   d. decreased blood glucose
   e. linear growth

7. Dwarfism may result from deficiency of:
   a. GH
   b. receptors for somatomedin
   c. IGF-I
   d. a and b
   e. a, b, and c

8. The first step in the synthesis of thyroid hormones is:
   a. iodination of tyrosine
   b. coupling of iodinated tyrosine to thyroglobulin
   c. coupling of iodinated tyrosines to form DIT
   d. synthesis of thyroglobulin
   e. release of T_3 and T_4

9. Under which condition would thyroid hormone be secreted?
   a. decreased TSH, decreased TRH
   b. increased TRH, decreased TSH
   c. increased TRH, increased TSH
   d. decreased TRH, increased TSH

10. The effects of thyroid hormones include all of the following except:
   a. calorigenesis
   b. increased cardiac output
   c. decreased lipid oxidation
   d. glycogen formation
   e. myelinization of nerves

VI. Fill in the Blanks. For each of the situations, indicate whether growth hormone increases, decreases, or is unaffected.

_________ 1. absence of somatotrophs

_________ 2. low blood glucose concentration

_________ 3. excess of somatostatin

_________ 4. destruction of posterior pituitary

_________ 5. high blood glucose concentration

_________ 6. cut through the median eminence

VII. Fill in the Blanks. For each situation indicate the effect (increase, decrease, none) on the secretion of thyroid hormone.

_________ 1. dietary deficiency of iodine

_________ 2. cut through the median eminence

_________ 3. destruction of the posterior pituitary

_________ 4. destruction of gonadotrophs

_________ 5. lack of TSH

_________ 6. excess of TRH

_________ 7. excess of thyroid hormones in the plasma

_________ 8. dietary deficiency of protein

_________ 9. decrease in body temperature

VIII. Complete the Table.

Hypothalamic hormone	Cell type in anterior pituitary	Hormone released from anterior pituitary	Target organ for anterior pituitary hormone	Hormone released by target organ
	Corticotrophs			
TRH				
		Prolactin		
			ovaries	
				testes

Chapter 14

The Adrenal Glands

Chapter Learning Objectives

1. Describe the gross anatomy of the adrenal glands, particularly focusing on the three zones of the adrenal cortex.
2. Indicate the factors that control the action of the adrenal cortex.
3. Summarize the results of aldosterone secretion.
4. Describe the chemical nature of the hormones of the adrenal cortex and medulla and indicate how they are synthesized.
5. Diagram negative feedback mechanisms in glucocorticoid secretion.
6. Indicate the characteristics of the diurnal secretion of cortisone and how the environment may modify its concentration.
7. Summarize the effects of glucocorticoid secretion throughout the various tissues of the body.
8. State the differences in the physiological and pharmacological actions of the steroids.
9. Explain the regulation of mineralocorticoid secretion and how the mineralocorticoids affect the fluid balance of the body; describe the feedback loop involving the kidney, potassium ions, sodium, and aldosterone in the regulation of mineralocorticoid secretion.
10. Compare the effects of Cushing's syndrome and Addison's disease.
11. Characterize the role of the adrenal medulla and the action of its principal hormone, epinephrine; describe the relationship between the adrenal medulla and sympathetic nervous system and then compare the effects of each on the body.

Expanded Chapter Outline

I. The Adrenal Cortex

A. *Cells of the adrenal cortex show three specific zones, and each zone produces a specific class of steroid hormone:*

1. **Zona glomerulosa,** which produces mineralocorticoids.
2. **Zona fasciculata,** which produces glucocorticoids.
3. **Zona reticularis,** which produces and secretes primarily androgens.

B. *Activity of the adrenal cortex is under control of ACTH produced by the pituitary.*

1. The release of **ACTH** is regulated by **CRH**.
2. ACTH promotes growth and function of all three zones and regulates glucocorticoid secretion by the cells of the zona fasciculata.

C. *Three steroids are the primary products of the adrenal cortex.*

1. The first step in steroid synthesis is the conversion of cholesterol to **pregnenolone**.
2. Steroid synthesis occurs on demand and is regulated by the availability of the enzyme catalyzing the first step.
3. Each zone in the cortex contains the enzymes for synthesizing its specific steroid hormone.

D. *Glucocorticoids play an important role in metabolic regulation.*

1. Secretion of glucocorticoid is regulated by a sequence of steps beginning with the release of CRH by the hypothalamus.
   a. The release of CRH is determined by inputs from higher brain centers to the hypothalamus.
   b. Glucocorticoid secretion is responsive to circadian variations in CRH and ACTH secretion, which are related to the individual's sleep-wake patterns.
   c. Glucocorticoid synthesis and secretion are responsive to physically or physiologically stressful stimuli, such as hypoglycemia, physical trauma, and

strenuous exercise, and by emotionally stressful stimuli, such as acute anxiety.

d. Glucocorticoid secretion is regulated by a series of negative feedback loops.

e. **Cortisol** is transported in the blood by carrier proteins.

f. Glucocorticoids complex with intracellular receptors and exert their effect on DNA.

g. Cortisol exerts its primary effects on the liver, skeletal muscle, and adipose tissue. The overall effect is **catabolic.**

h. Cortisol has an amplifying effect on the action of other hormones and enhances the responsiveness of blood vessels.

2. The two most important pharmacological effects of glucocorticoids are:
   a. **Anti-inflammatory effects.**
   b. **Immunosuppressive effects.**

E. Mineralocorticoids play an important role in regulating mineral metabolism.

1. The two primary regulators of aldosterone secretion are:
   a. Potassium ions.
   b. **Angiotensin II,** a peptide hormone, formed by a sequence of steps that begin in the kidney and are collectively known as the **renin-angiotensin system:**
      i. **Renin,** produced by the kidney, acts on **angiotensinogen** to form **angiotensin I.**
      ii. In the lungs this molecule is converted to **angiotensin II,** which then travels to the adrenal cortex.
      iii. The amount of aldosterone is related to the amount of renin released from the kidney.

2. An increase in plasma potassium concentration stimulates aldosterone secretion.
   a. Aldosterone promotes potassium excretion by the kidneys.
   b. Aldosterone and the kidneys comprise two components of a feedback loop regulating plasma potassium concentration.

3. Aldosterone plays an important role in regulating fluid balance in the body and acts primarily on the kidneys.

F. Abnormalities of adrenal cortex function lead to altered physiological regulation.

1. Cushing's syndrome results from excess cortisol secretion.

2. Addison's disease results from an overall deficiency in steroid production by the adrenal cortex.

II. The Adrenal Medulla

A. The medulla is centrally located in each adrenal gland.

1. The adrenal medulla is an important component of the sympathetic nervous system and secretes catecholamines.

2. The functional unit of the **adrenal medulla** is the **chromaffin cell,** which is innervated directly by preganglionic fibers of the sympathetic nervous systems.

3. Chromaffin cells secrete **epinephrine,** which is carried via the blood to various effector organs in the body.

B. The primary hormone secreted by the adrenal medulla is epinephrine.

1. The pathway for synthesis of epinephrine is:

tyrosine → DOPA → norepinephrine → epinephrine

The rate limiting step is the conversion of tyrosine to DOPA.

2. Epinephrine is packaged and concentrated in vesicles known as **chromaffin granules,** which are released when the sympathetic neuron discharges.

C. Epinephrine triggers a sequence of responses appropriate to "fight-or-flight" situations.

1. The adrenal medulla triggers a sequence of events appropriate to "fight-or-flight" situations.

2. Many effects serve to increase and redistribute blood flow between various tissues.

3. Epinephrine increases the ability of the blood to:
   a. Transport oxygen.
   b. Affect metabolic processes in several tissues (liver, skeletal muscle, and adipose tissue).

c. Increase the breakdown of stored fuels into metabolizable substrates.

Key Terms

Addison's disease
adrenal cortex
adrenal medulla
adrenocorticotropic hormone (ACTH)
aldosterone
androgen
angiotensin I, II
angiotensin-converting enzyme
angiotensinogen
anti-inflammatory
catabolic
chromaffin cell
chromaffin granule
circadian rhythm
corticosteroid binding globulin
cortisol norepinephrine
Cushing's syndrome
dihydroxyphenyl-alanine (DOPA)
epinephrine
fight-or-flight
glucocorticoid
gluconeogenesis
granular cells
hypothalamic-pituitary-adrenal axis
immunosuppressive
lipolysis
mineralocorticoid
negative feedback
permissive action
renin
renin-angiotensin system
steroidogenic effect
trophic effect
tyrosine hydroxylase
vascular reactivity
zona fasciculata
zona glomerulosa
zona reticularis

Testing Your Knowledge

I. Matching. Select the correct term from the Key Terms.

________ 1. These cells produce glucocorticoids

________ 2. These cells produce mineralocor ticoids

________ 3. These cells produce androgens

________ 4. site of production of epinephrine

________ 5. caused by hypersecretion of glucocorticoids

________ 6. caused by hyposecretion of glucocorticoids and mineralocorticoids

________ 7. production affected by sleep-wake pattern and circadian variation

________ 8. pharmacological effect of high concentration of cortisol

________ 9. immunosuppressive hormone produced from cholesterol

________ 10. role in regulation of mineralocorticoid secretion, produced by kidney

________ 11. end product of renin-angiotensin system that binds to receptors in cortex

________ 12. increase in plasma K^+ concentration causes increase of this hormone

________ 13. hormone synthesized from tyrosine

________ 14. stimulates lipolysis in adipose tissue

________ 15. transports cortisol in blood

II. Dichotomous. Circle the correct choice from the paired terms in parentheses.

1. Glucocorticoids are produced in the zona (glomerulosa/fasciculata).
2. ACTH regulates secretion of hormones from the adrenal (cortex/medulla).
3. Cortisol secretion would (increase/decrease) in a person with hypoglycemia.
4. As cortisol level increases in the blood, CRH (increases/decreases).
5. Glucocorticoids and mineralocorticoids bind to receptors located (on the surface of/inside the) cell.
6. The overall effect of cortisol is (anabolic/catabolic).
7. Glucocorticoids tend to (increase/decrease) blood glucose.
8. In adipose tissue and skeletal muscle, the effect of glucocorticoids is (the same as/opposite to) that of insulin.
9. A(n) (increase/decrease) of plasma K^+ stimulates aldosterone secretion and (raises/lowers) plasma K^+ concentration.
10. The net effect of epinephrine metabolism is (increased/decreased) blood glucose concentration.

III. True / False. In the space to the left of the number write T for true or F for false.

_______ 1. Androgens are produced by the adrenal medulla from cholesterol.

_______ 2. Cholesterol cannot be stored in cells and must be actively transported from the blood.

_______ 3. Cortisol regulates its own production via ACTH and CRH.

_______ 4. Most cortisol is transported as a free, unbound molecule in the blood.

_______ 5. Cortisol exerts its effects through G proteins.

_______ 6. Cortisol promotes loss of protein in skeletal muscle.

_______ 7. Glucocorticoids enhance the activity of the immune system.

_______ 8. The major site of action of aldosterone is the liver.

_______ 9. A decrease in blood flow to the kidney inhibits release of renin.

_______ 10. Cortisol enhances synthesis of nor epinephrine.

IV. Sentence Completion. Write your response in the space provided.

1. The components of the hypothalamic-pituitary-adrenal axis are _________, _________, _____________ and the adrenal.

2. The rate limiting step in steroid synthesis is the conversion of _________ to _________.

3. The circadian rhythm for glucocorticoid secretion occurs approximately every ___________ hours.

4. In the liver, glucocorticoids promote the formation of glucose by _______________.

5. In the skeletal muscle, protein synthesis is _________________.

6. During stress, cortisol treatment restores and enhances responsiveness of blood vessels of _________________.

7. High doses of glucocorticoids decrease the number of __________, __________, and ______________ in the blood, but increase ______________, ______________, and ______________.

8. The two primary regulators of aldosterone secretion are __________ and __________.

9. Cushing's syndrome results from ________ secretion of glucocorticoids, but Addison's disease results from a _______________.

10. The primary hormone secreted by the adrenal medulla is _________________.

V. Multiple Choice. Circle the letter of the best response.

1. Which of the following pairs is mismatched?
   a. zona glomerulosa — mineralocorticoids
   b. zona fasciculata — glucocorticoids
   c. zona reticularis — androgens
   d. chromaffin cell — epinephrine
   e. none is mismatched

2. In the absence of ACTH, secretion of which of the following would be affected the most?
   a. cortisol
   b. aldosterone
   c. dehydroepiandrosterone
   d. epinephrine
   e. angiotensin I

3. Over a 24-hour period in a normal individual, plasma cortisol levels are:
   a. highest at noon and lowest at midnight
   b. highest around time of awakening and lowest at midnight
   c. highest at midnight and lowest at noon
   d. highest at noon and midnight
   e. lowest around the time of awakening and at midnight

4. Under which condition would cortisol be secreted?
   a. fasting
   b. breaking a bone
   c. preparation for a race
   d. impending final examinations
   e. all of the above

5. Secretion of cortisol is regulated by negative

feedback loops in which cortisol:
a. inhibits secretion of ACTH
b. increases sensitivity of corticotrophs to CRH
c. increases secretion of CRH
d. stimulates ACTH to inhibit the hypothalamus
e. decreases synthesis of CRH but increases ACTH synthesis

6. Cortisol exerts its primary effects on:
a. liver
b. skeletal muscle
c. adipose tissue
d. a and b
e. a, b, and c

7. Which of the following is <u>not</u> an effect of cortisol in metabolism?
a. gluconeogenesis
b. increased glycogen synthesis
c. increased protein synthesis in skeletal muscle
d. decreased glucose uptake in skeletal muscle and adipose tissue
e. increased lipid mobilization

8. In the renin-angiotensin system, the first event to occur is:
a. angiotensinogen → angiotensin I in blood
b. angiotensin I → angiotensin II in lungs
c. angiotensin II → aldosterone synthesis
d. secretion of renin
e. secretion of angiotensinogen

9. The rate limiting step in synthesis and secretion of epinephrine is:
a. conversion of tyrosine → DOPA
b. packaging of epinephrine into chromaffin granules
c. conversion of DOPA → norepinephrine
d. conversion of norepinephrine → epinephrine
e. none of the above

10. All of the following are effects of epinephrine except:
a. increased cardiac output
b. vasoconstriction of skeletal muscles
c. increased glycogenolysis in liver
d. increased glucagon secretion
e. decreased insulin secretion

VI. Fill in the Blanks. For each situation, indicate the effect (increase, decrease, none) on the production of a glucocorticoid-like cortisol.

__________ 1. cut through median eminence

__________ 2. destruction of posterior pituitary

__________ 3. destruction of zona glomerulosa

__________ 4. hypoglycemia

__________ 5. increased plasma concentration of cortisol

__________ 6. decrease in level of ACTH

__________ 7. increase in CRH

VII. Fill in the Blanks. For each situation, indicate the effect (increase, decrease, none) on the production of aldosterone.

__________ 1. destruction of anterior pituitary

__________ 2. destruction of zona reticularis

__________ 3. lack of ACTH

__________ 4. deficiency of renin

__________ 5. deficiency of receptors to angiotensin II

__________ 6. increase in blood flow to kidney

__________ 7. decrease in plasma potassium ion concentration

VIII. Fill in the Blanks. For each situation, indicate the effect (increase, decrease, none) on the production of epinephrine by the adrenal medulla.

__________ 1. deficiency of tyrosine hydroxylase

__________ 2. stimulation of sympathetic nervous system

__________ 3. being chased by an angry dog

__________ 4. destruction of the posterior pituitary

__________ 5. cut in the hypothalamic-pituitary portal system

Chapter 15

The Endocrine Pancreas

Chapter Learning Objectives

1. Distinguish between the endocrine and exocrine cells of the pancreas, and detail the hormones produced by the islets of Langerhans.
2. Indicate how different stimuli affect insulin secretion.
3. Indicate how different stimuli affect glucagon secretion.
4. Describe the metabolic effects of insulin on carbohydrate, lipid, and protein metabolism.
5. Describe the metabolic effects of glucagon on carbohydrate, lipid, and protein metabolism.
6. Give the significance of the insulin/glucagon ratio, particularly to metabolic activities.
7. Indicate how hormones regulate metabolism during fed and fasting states.
8. Describe the complications of diabetes mellitus.
9. Compare type I and type II diabetes.
10. Distinguish between acute and chronic effects of diabetes.

Expanded Chapter Outline

I. Islets of Langerhans: Functional Units of the Endocrine Pancreas

A. The major cell types of the endocrine pancreas are:

1. **Alpha cells**, which produce glucagon.
2. **Beta cells**, which produce insulin.
3. **Delta cells**, which produce somatostatin.
4. **F cells**, which produce pancreatic polypeptide.
5. Each islet shows an orderly arrangement of the different types of cells.
6. Islets are innervated by the sympathetic and parasympathetic divisions of the autonomic nervous system.

II. Synthesis and Secretion of Pancreatic Hormones

A. Insulin is secreted in response to increased glucose in the blood.

1. **Insulin** is derived from the precursor, proinsulin.
   a. Proinsulin consists of a single peptide chain containing insulin subunits A and B linked together by a third peptide.
   b. As proinsulin is packaged into secretory vesicles, enzymes cleave **proinsulin** into the **A and B chains** and the **C-peptide**, whose only function appears to be its role in insulin synthesis.

2. Several factors stimulate the release of insulin, which is stored in secretory vesicles in the beta cells. These include:
   a. Increase in blood glucose concentration.
   b. Ingestion of high-protein meat and amino acids such as arginine.
   c. Presence of fatty acids.
   d. Gastrointestinal hormones such as **gastrin** and **secretin**.
   e. Acetylcholine from postganglionic fibers of the parasympathetic nervous system.
   f. Drugs such as sulfonylurea.
   g. Other hormones, such as glucagon.

3. Inhibitors of insulin secretion include:
   a. Somatostatin
   b. Epinephrine
   c. Norepinephrine

B. Glucagon is secreted in response to decreased glucose in the blood.

1. **Glucagon** is synthesized by **alpha cells** of the pancreas and derived from a precursor, **proglucagon**, which is converted to the active hormone.

2. Factors stimulating glucagon secretion include:
   a. Decrease in blood glucose concentration.
   b. Amino acids such as arginine.
   c. Acetylcholine and epinephrine and norepinephrine from the parasympathetic and sympathetic divisions of the autonomic nervous system.

3. Inhibitors of glucagon secretion include:
   a. Increase in circulatory fatty acids.
   b. Insulin.
   c. Somatostatin.

C. Somatostatin is secreted in response to increased glucose and glucagon in the blood.

1. **Somatostatin** is derived from the precursor, prosomatostatin.

2. Factors stimulating secretion of somatostatin from delta cells include:
   a. Increased blood glucose.
   b. Increased plasma glucagon.
   c. Amino acids.

3. Secretion of insulin, glucagon, and somatostatin is interrelated.

III. The Metabolic Effects of Pancreatic Hormones

A. Glucagon and insulin serve as key components in feedback loops, which regulate full homeostasis.

B. Tight regulation of blood glucose concentration is necessary to ensure that the central nervous system, which relies almost solely on glucose for its metabolic needs, has an adequate supply.

C. Insulin has anabolic effects on metabolism.

1. The major target tissues of insulin are skeletal muscle, adipose tissue (fat), and liver.
   a. Brain and liver cells are readily permeable to glucose, even in the absence of insulin.
   b. Glucose enters other tissues such as muscle and adipose by both simple and facilitated diffusion.
   c. Insulin promotes the movement of glucose transporters from their intracellular location to the plasma membrane.

2. Effects of insulin on carbohydrate metabolism include:
   a. Stimulation of glycogenesis in liver and muscle.
   b. Stimulation of enzymes of the glycolytic pathway.
   c. Decreased glucose output by the liver.

3. Effects of insulin on lipid metabolism include:
   a. Promotion of the storage of excess fuel as triglycerides.
   b. Promotion of the synthesis of both fatty acid and glycerol-phosphate components of triglycerides.
   c. Inhibition of lipolysis or triglyceride breakdown.

4. Effects of insulin on protein metabolism include:
   a. Protein accumulation in muscle, liver, and adipose tissue by increased protein synthesis and decreased protein degradation.
   b. Production of a **"positive nitrogen balance."**

B. Glucagon has effects on metabolism that oppose those of insulin.

1. Effects of glucagon are mediated through second messengers.

2. Glucagon acts principally on the liver, and its overall effect on metabolism is **catabolic**.

3. Glucagon increases blood glucose concentration by stimulating **glycogenolysis** and **gluconeogenesis**.

4. Glucagon stimulates **lipolysis** and **ketogenesis** in the liver.

5. Glucagon stimulates **protein degradation** in the liver and the transport of amino acids from the blood to the liver for gluconeogenesis and urea synthesis.

C. The insulin/glucagon ratio in the blood is an important determinant of metabolic flow.

1. The **insulin/glucagon (I/G)** ratio is probably more important in determining the overall metabolic state than is the absolute concentration of either hormone.

2. When the I/G ratio is high, the effects of insulin predominate, and an anabolic state exists.

3. When the I/G ratio is low, the effects of glucagon predominate, and a catabolic state exists.

IV. Overview of the Metabolic Regulation by Pancreatic Hormones: The Fed State Versus the Fasted State

A. *Excess fuel is stored in the body in several forms.*

1. Most calories in the body are stored as triglycerides in adipose tissue. Muscle protein represents about 15% of total stored fuels.

2. Few calories are stored in glycogen in the muscle and liver.

B. *Flow of nutrients after a meal is largely directed by insulin.*

1. In the fed state, the combination of high insulin and low glucagon promotes fuel storage in liver, adipose, and muscle.

2. Glycogen storage is increased in liver and muscle, triglyceride storage is increased in adipose tissue, and protein synthesis is promoted in all three tissues.

C. *Glucagon influences the flow of nutrients during fasting.*

1. In the fasted state, glycogen stores are depleted, muscle protein is degraded, and amino acids are used for gluconeogenesis in the liver.

2. The triglyceride stores in adipose tissue are broken down.

V. Diabetes Mellitus

A. *Diabetes mellitus is a heterogeneous disease characterized by high blood glucose concentration.*

~ The hyperglycemia results from a deficiency, either relative or absolute, in insulin action.

~ Two forms of diabetes make up 98% of all cases. In both forms, there is a genetic component.

1. Type I, Insulin-dependent diabetes, mellitus, results from an inability of the pancreas to produce adequate amounts of insulin.

2. Type II, noninsulin-dependent diabetes mellitus, results from an inability of target tissues to respond to insulin.

B. *Diabetes mellitus can result in both short-term and long-term complications.*

1. Acute complications of diabetes include **hyperglycemia**, **polyurea**, and **dehydration**, as well as **ketoacidosis** and electrolyte imbalances in type I.

2. Chronic secondary complications of diabetes include:
   a. Reduced circulation to the brain, heart, and lower extremities resulting in stroke, heart attack, or loss of limb.
   b. Lesions in microvasculature in the kidney and retina, which may lead to kidney disease and diabetic retinopathy.
   c. Diabetic neuropathy involving fibers of the autonomic nervous system and peripheral sensory nerves.

Key Terms

alpha cells
anabolic effect
beta cells
C-peptide
calorie density
catabolic effect
pancreatic polypeptide
dehydration
delta cells
diabetic neuropathy
diabetic retinopathy
electrolyte imbalance
endocrine pancreas
exocrine pancreas
F cells
gestational diabetes
glucagon
gluconeogenesis
glucosuria
glycogenolysis
hormone-sensitive lipase
insulin
insulin-dependent diabetes mellitus (type I)
insulin/glucagon ratio (I/G)
insulin resistance
islets of Langerhans
ketoacidosis
ketogenesis
ketone
ketone bodies
lipoprotein lipase
nitrogen balance
noninsulin-dependent diabetes mellitus (type II)
hyperglycemia
polyuria
proglucagon
proinsulin
prosomatostatin
somatostatin
sulfonylurea
triglyceride
urea synthesis

Testing Your Knowledge

I. Matching. Select the correct term from the Key Terms.

________ 1. pancreatic cells producing glucagon

________ 2. pancreatic cells producing insulin

________ 3. groups of cells comprising endocrine pancreas

________ 4. precursor of insulin and C-peptide

________ 5. secreted in response to low blood glucose

________ 6. secreted in response to high blood glucose

________ 7. general effect of insulin

________ 8. conversion of glycogen to glucose

________ 9. conversion of nonglucose substrates into glucose

________ 10. disease condition caused by loss of beta cell function

________ 11. disease condition caused by relative lack of insulin action on its target tissues

________ 12. describe elevated blood glucose level

________ 13. secondary effect resulting from urinary loss of ketones

________ 14. results in loss of feeling in lower limbs

________ 15. pancreatic cells producing somatostatin

II. Dichotomous. Circle the correct choice from the paired terms in parentheses.

1. The (exocrine/endocrine) part of the pancreas secretes insulin and glucagon.
2. Insulin is secreted when the blood concentration of glucose (increases/decreases).
3. (Epinephrine/Acetylcholine) inhibits secretion of insulin.
4. Glucagon is secreted by the (alpha/beta) cells of the islets.
5. During periods of fasting, the concentration of (insulin/glucagon) would be highest.
6. Insulin has (anabolic/catabolic) effects on metabolism.
7. In the presence of insulin, protein synthesis (increases/decreases) in muscle.
8. The conversion of glycogen into glucose is known as (gluconeogenesis/glycogenolysis).
9. When the I/G ratio is high, the effects of (insulin/glucagon) predominate.
10. During fasting, blood levels of fatty acid and ketone (increase/decrease).

III. True/False. In the space to the left of the number write T for true or F for false.

______ 1. Glucagon is secreted by the alpha cells in the islets.

______ 2. As the concentration of glucose in the blood increases, the secretion of insulin decreases.

______ 3. After a meal rich in protein, but low in carbohydrate, both insulin and glucagon are secreted.

______ 4. Somatostatin is secreted by beta cells.

______ 5. The receptor for insulin is located in the cytoplasm of the target cell.

______ 6. In the presence of insulin, lipolysis is decreased.

______ 7. The cellular effects of glucagon are mediated by cAMP.

______ 8. The principal target of glucagon is the liver.

______ 9. When the I/G ratio is low, the effects of insulin predominate.

_______ 10. During fasting, the brain may use amino acids as a source to make glucose.

IV. Sentence Completion. Write your response in the space provided.

1. __________ is secreted by the beta cells of the islets.

2. __________ provides an indirect measure of insulin production.

3. The ____________ relies primarily on glucose for its metabolic needs.

4. In a normal individual, the blood glucose concentration following an overnight fast is in the range of __________________ mg of glucose per 100 mL of blood.

5. Insulin stimulates the activity of __________ to promote uptake of lipoproteins from the blood.

6. In muscle and liver, insulin ______________ protein synthesis and ____________ protein degradation.

7. The principal target site for glucagon is the _______________, where its overall effect in metabolism is_______________.

8. Glucagon stimulates the formation of __________________ in the liver and heart thereby sparing glucose.

9. When the I/G ratio is ____________, excess fuels are stored as glycogen and triglycerides.

10. The calorie density of ____________ is twice that for _________________.

V. Multiple Choice. Circle the letter of the best response.

1. Which of the following pairs is mismatched?
   a. alpha cells — glucagon
   b. beta cells — insulin
   c. delta cells — somatostatin
   d. gamma cells — somatomedin
   e. F cells — pancreatic polypeptide

2. All of the following are derived from proinsulin except:
   a. A chain
   b. B chain
   c. C peptide
   d. d chain
   e. a and b

3. Insulin secretion is stimulated by all of the following except:
   a. epinephrine
   b. acetylcholine
   c. glucagon
   d. gastrin
   e. increased blood glucose

4. Glucagon secretion is stimulated by all of the following except:
   a. low blood glucose
   b. acetylcholine
   c. insulin
   d. epinephrine
   e. arginine

5. The major target tissues of insulin are:
   a. skeletal muscle
   b. fat
   c. brain
   d. a and b
   e. a, b, and c

6. Glucose enters cells primarily by the mechanism of:
   a. facilitated diffusion
   b. endocytosis
   c. exocytosis
   d. active transport
   e. diffusion

7. All of the following are effects of insulin on carbohydrate metabolism except:
   a. increased glycogen synthesis
   b. increased glycolysis
   c. increased gluconeogenesis
   d. increased use of amino acid
   e. all are effects

8. When insulin is present, all of the following occur except:
   a. protein synthesis
   b. fatty acid synthesis
   c. stimulation of lipoprotein lipase
   d. lipolysis
   e. glycogenesis

9. Which does not occur in the presence of glucagon?
   a. fatty acid synthesis
   b. glycogenolysis

c. gluconeogenesis
d. lipolysis
e. ketogenesis

10. Which of the following pairs is mismatched?
   a. fed state — I/G ratio low
   b. fed state — glycogenesis increased
   c. fed state — stimulation of lipoprotein lipase
   d. fasted state — stimulation of gluconeogenesis
   e. fasted state — stimulation of ketogenesis

11. Which of the acute complications of diabetes would probably occur first?
   a. ketoacidosis
   b. electrolyte imbalance
   c. polyuria
   d. hyperglycemia
   e. dehydration

VI. Complete the Table.

Hormone	Source	Stimulators	Inhibitors	Effect on carbohydrate metabolism	Effect on lipid metabolism	Effect on protein metabolism
Insulin						
Glucagon						

VII. Fill in the Blanks. For each situation, indicate the effect (increase, decrease, none) on insulin secretion.

__________ 1. decrease in blood glucose level

__________ 2. ingestion of meal high in proteins

__________ 3. increased secretion of somatostatin

__________ 4. destruction of anterior pituitary

__________ 5. activation of sympathetic nervous system

__________ 6. destruction of alpha cells

VIII. Fill in the Blanks. For each situation, indicate the effect (increase, decrease, none) on glucagon secretion.

__________ 1. increase in blood glucose level

__________ 2. ingestion of high-protein meal

__________ 3. decrease in fatty acid concentration

__________ 4. activation of parasympathetic nervous system

__________ 5. destruction of beta cells

__________ 6. destruction of anterior pituitary

Chapter 16

Muscles

Chapter Learning Objectives

1. Consider a muscle as an effector, a motor, and a regulator.
2. Compare the different muscle types and indicate their most important distinctions in structure, function, innervation, and regulation of contraction.
3. Indicate how an action potential is propagated throughout a muscle and into a muscle fiber.
4. Explain the role of calcium in muscle contraction.
5. Describe the different qualities of muscle contraction, including duration, force, velocity, and power.
6. Compare the length-tension and force-velocity curves, regarding the type of relationship represented by each curve.
7. Evaluate the differences in isotonic and isometric contraction.
8. Describe the structure of a sarcomere, particularly the relationship among the sarcolemma, t-tubules, and sarcoplasmic reticulum (SR).
9. Give the molecular explanation for the sliding filament theory of contraction.
10. Consider how certain chemicals affect neural transmission.
11. Compare white, slow and fast, red muscle fibers as to contraction speed, resistance to fatigue, ATPase activity, myoglobin content, number of mitochondria, vascularity, and function in the body.

Expanded Chapter Outline

I. Overview: The Roles and Types of Muscle

A. *Muscle serves as the output for the central nervous system.*

1. Muscle effects changes prescribed by the nervous system.

2. Muscle also helps to regulate input into the nervous system by adjusting the sensitivity of nerve organs.

B. *Muscle is a biological motor.*

1. Muscles consume fuel, do useful work, and produce heat.

2. Muscle plays a role in conscious "cooling" mechanisms.

C. *Muscle aids in regulating physiological processes.*

1. Muscles control movement of substances through tubular structures.

2. They play a central role in maintenance of posture and control of body temperature.

D. *Muscle can be classified in several ways.*

1. Muscles may be categorized by location and function.
   a. **Skeletal** muscles attached to bones are used for locomotion and maintaining body posture.
   b. **Visceral** muscles line the walls of internal organs and are also located in blood vessels and sensory organs.
   c. **Cardiac** muscle is found only in the heart.

2. Muscles may be classified by structure.
   a. Skeletal and cardiac muscles appear striated on viewing with a microscope.
   b. Smooth muscle, e.g., visceral, is not striated.

3. Muscles may be classified by mode of action and control.
   a. Skeletal muscle must receive a specific signal from the central nervous system in order to function.
   b. Most movements of skeletal muscle are voluntary.

c. The actions of most smooth muscle are involuntary and initiated by the autonomic nervous system.
d. Cardiac muscle is involuntary but has an intrinsic rate and strength of contraction.
e. Smooth muscle may be classified as **multiunit** or **unitary,** depending on its mode of control.

II. The Structure of Muscle

A. Skeletal muscle structure and function are closely related.

1. The skeleton works with the muscles as a lever system.
   a. Muscles attach to bones by **tendons**.
   b. Skeletal muscle anatomy reflects adaptation to specific functions.
   c. The range of motion of a muscle may be limited by the skeletal system.
   d. Muscles are arranged as **antagonistic pairs**, e.g., flexor and extensor.

2. A muscle is composed of muscle fibers (cells) that are covered by a connective tissue sheet.
   a. Muscle is highly vascularized.
   b. Tendons are extensions of the connective tissue sheets.

3. The striations seen in muscle fibers are composed of myofilaments, which differ from each other in their size and molecular composition.
   a. **Myofilaments** are organized into a sarcomere, the basic unit of contraction in the skeletal muscle.
   b. The thick myofilaments composed of **myosin** and light chains overlap the thin filaments composed of **actin**, a **troponin** complex of three types of troponin and **tropomyosin**.
   c. Other distinguished areas of a **sarcomere** are the A-band, I-band, Z-line, H-zone, and **crossbridges**.
   d. Accessory proteins are associated with myofilaments.
   e. Hundreds of sarcomeres connected in series make up a **myofibril** running the length of a muscle fiber cell.

4. There are two membrane systems in a muscle fiber cell.
   a. Extensions of the **sarcolemma**, the plasma membrane of a muscle cell, extend inward as **t-tubules**.
   b. The **sarcoplasmic reticulum** contacts the t-tubules at the **triad** in skeletal and cardiac muscle cells and at the **dyad** in smooth muscle cells.

B. Cardiac muscle structure is similar to that of skeletal muscle.

1. **Desmosomes** and **gap junctions** provide a strong mechanical and electrical connection, respectively, among cardiac muscle cells.

2. The intracellular organization of cardiac and skeletal muscle cells is similar.

C. Smooth muscle is made of small cells without striations.

1. Smooth muscle has a different cellular and tissue structure.
   a. Cells form gap junctions, which allow for coordination of smooth muscle activity.
   b. A network of collagen and elastic fibers surrounds and links cells together and provides mechanical strength.

2. Smooth muscle is organized into layers: circular and longitudinal.
   a. Coordinated movement often involves the alternate contraction and relaxation of two muscle layers.
   b. Blood vessels have just the circular muscle layer.

III. The Function of Muscle

A. Skeletal muscle contraction depends on external conditions.

1. **Isometric contractions** in the laboratory provide clues to the sliding-filament hypothesis of contraction.
   a. In an isometric contraction, force is developed, but no shortening occurs.
   b. The length of a muscle before contraction has an important effect on the amount of force the muscle can exert when it is stimulated.
   c. If a muscle is progressively stretched between contractions, the peak amount of contraction force increases until the length at which twitch force is greatest.
   d. At this twitch point, the active force decreases relative to the passive or resting force.

e. The isometric **length-tension curve** describes the relationship in which the force capability of the muscle is greatest at some intermediate length.
f. Maximum force capability occurs at an optimal length, L_O, and peak force at this length is P_O or F_O.
g. The amount of force developed is directly related to the amount of myofilament overlap at lengths greater than L_O.

2. **Isotonic contractions** provide clues to the work of muscle as a motor.
   a. In an isotonic contraction, the force in the muscle remains constant as the muscle shortens.
   b. A mixed contraction consisting of an initial isometric contraction followed by an isotonic shortening occurs in an afterloaded muscle.
   c. The weight of the **afterload** affects the length of the isometric portion of the mixed contraction.
   d. The weight of the afterload directly affects the speed at which a muscle shortens and the degree of shortening.
   e. This relationship between afterload weight and speed of contraction is described by the **force-velocity curve**.
   f. $\mathbf{V_{max}}$, the highest velocity of shortening, provides an indication of the maximal rate of energy-converting processes in a muscle.
   g. The **power output** of the muscle varies with the afterload and is zero at maximum force.

B. Cardiac and smooth muscle operate under specialized mechanical conditions.

1. The contraction of cardiac and smooth muscle can be described by length-tension and force-velocity curves.
   a. Cardiac muscle operates at lengths less than L_O.
   b. Contraction of cardiac muscle is mainly isometric.

2. Differences between different types of muscle are found in the speed of contraction and shape of the length-tension curve.

IV. The Control of Muscle

A. Cell membrane activity is important in controlling muscle function.

1. Stimulation of the muscle membrane generates an action potential.
   a. The action potential is caused by time-dependent and voltage-dependent changes in the membrane's permeability to sodium and potassium.
   b. Muscle cells are stimulated by way of motor neurons.

2. The action potential is conducted into the depths of the muscle fiber through the transverse tubular system (t-system).

3. The t-tubules are in close contact with membranes of the sarcoplasmic reticulum (SR) that stores and releases ionized calcium, the substance controlling contractile activity of muscle proteins.
   a. Ca^{2+} is stored in terminal cisternae of the SR.
   b. Each sarcomere has its own portion of the SR around myofilaments.

B. Calcium ions are a vital link in the control of muscles.

1. The sequence of events during muscle contraction is:
   a. Generation of action potential through t-tubules.
   b. Release of Ca^{2+} from SR.
   c. Binding of calcium ions to troponin complex attached to tropomyosin on action.
   d. Change in conformation of the troponin complex and in the position of tropomyosin on actin.
   e. Attachment of actin with myosin cross bridge.
   f. Movement of actin filaments toward center of the sarcomere by myosin head in power stroke.

2. Repetition of the cycle of formation of cross bridges between actin and myosin, power stroke, and detachment of myosin from actin leads to a shortening of the sarcomere.
   a. ATP provides energy for power stroke of myosin head and for detachment of myosin head from actin.
   b. The cycle is repeated as long as ATP and Ca^{2+} are available.

3. When stimulation of the muscle ceases, Ca^{2+} is pumped back into the SR, troponin returns to its precontraction conformation,

and tropomyosin resumes its blocking function on actin.

4. Skeletal and cardiac muscle contraction is regulated through an action on the thin filaments, called **actin-linked regulation**.
   a. Maximal contraction is not attained in cardiac muscle because of an insufficient amount of Ca^{2+} released.
   b. Force of contraction of cardiac muscle may be increased by an increase in the duration or frequency of action potentials or the amount of Ca^{2+} entering the cell during an action potential.

5. Regulation of muscle contraction may be myosin-linked.
   a. Smooth muscle contraction is regulated through myosin filaments whose heads are attached to two protein chains called "light" chains.
   b. Attachment of phosphate ion to the myosin light chain activates the myosin chain and enables it to bind to actin.
   c. Actin filaments in smooth muscle lack the troponin protein complex.

6. Activation and regulation of smooth muscle can be divided into three groups of reactions occurring more or less simultaneously.
   a. Calcium ions entering the cytoplasm from the SR or extracellularly activate **calmodulin**, which is associated with **myosin light-chain kinase (MLCK)**.
   b. The kinase phosphorylates the regulatory myosin light chain, thereby activating it and allowing the binding to actin.
   c. A phosphatase removes the phosphate from the myosin.

7. The crossbridge cycle (attachment, rotation, detachment) to produce force and movement is similar to that occurring in skeletal muscle and continues as long as ATP is available and myosin is activated.

8. Relaxation occurs when Ca^{2+} is pumped back into the SR or removed from the cell. MLCK becomes inactivated, and phosphatase dephosphorylates myosin to prevent interaction with actin.

9. Smooth muscle contains a second regulatory process that enables it to maintain a contraction without consuming large amounts of energy.
   a. Modified crossbridges called "latch bridges" detach slowly and thus spend more time in a force-generating state.
   b. Control over contraction and relaxation is exercised by Ca^{2+}.

C. *The pattern of stimulation regulates muscle function.*

1. Summation of action potentials produces a sustained contraction in skeletal muscle.
   a. Stimulation of a muscle before relaxation is complete can result in a contraction that lasts longer and produces more than a single twitch. This process is called summation.
   b. The continuous supply of Ca^{2+} keeps the muscle from relaxing completely and produces a complete or fixed **tetanus**.

2. Distancing of action potentials produces discrete twitches in cardiac muscle.
   a. Cardiac muscle is a **twitch**-only type. In contrast to skeletal muscle, the action potential lasts as long as the contraction itself.
   b. Contractions can be adjusted by the amount of calcium made available for each contraction and by changes in muscle length.
   c. **Inotropic agents** (e.g., digitalis, calcium blockers, and epinephrine) permit changes in contractility of cardiac muscle.

3. **Phasic and tonic contractions** allow for a variety of responses in smooth muscle.
   a. Some smooth muscles give twitch-like or phasic contractions in response to a single stimulus or brief bursts of stimuli.
   b. Tonic contractions are long and sustained and develop in response to single or continued stimulation by neurons, drugs, or hormones.

D. *The central nervous system is important in regulating muscle function.*

1. Skeletal muscle contraction depends on nervous innervation from the somatic division of the central nervous system.
   a. **Acetylcholine** (ACh) is released from the motor neuron into the **myoneuronal junction**.
   b. All muscle fibers innervated by axons from the same motor neuron are activated together as part of a **motor unit**.

c. The muscle plasma membrane in the myoneuronal junction is folded and contains receptors for ACh and **acetylcholinesterase**.
d. The receptor's dual functions are binding to ACh and serving as an ion channel for Na and K.
e. During the action potential, the **end-plate potential** is about minus 15mV and is propagated down the muscle fiber to activate contraction.

2. Nervous control of smooth and cardiac muscle involves the autonomic nervous system (ANS).
   a. Multiunit smooth muscles are diversely innervated by autonomic nerve fibers, and each cell has a rudimentary motor end plate. Myoneuronal transmission is similar to that in skeletal muscles.
   b. In unitary smooth muscles, most cells are not innervated and instead receive their stimulus via local current flow, which occurs through gap junctions. The muscle acts as a functional **syncytium**.
   c. Autonomic nervous system innervation to cardiac muscle functions to regulate the timing and strength of contractions and not to initiate them.

3. Many factors block myoneuronal transmission and include:
   a. Too few Ca^{2+} or too many Mg^{2+}.
   b. Drugs interfering with reuptake of choline.
   c. Toxic molecules interfering with ACh release.
   d. Postsynaptic blocking agents that bind to ACh.
   e. Agents that inhibit acetylcholinesterase.
   f. Fatigue caused by depletion of ACh.
   g. Reduction in number of ACh receptors.

V. The Metabolism of Muscle Contraction

A. Energy for contraction comes from several sources.

1. Muscle cells use ATP for contraction and must have an adequate, constant supply for this process.

2. **Creatine phosphate**, a high-energy compound, serves as a reserve "energy pool" and transfers its phosphate to ADP to form ATP. Creatine is phosphorylated by ATP derived from cellular metabolism.

3. Glycolysis yields a small amount of cellular ATP.

4. Most ATP is produced via aerobic reactions in the mitochondria and yields 36 molecules of ATP from a glucose molecule.

5. Under ordinary activity, blood and oxygen supplies are adequate to allow for aerobic metabolism to provide sufficient ATP.
   a. When the level of activity reaches about 70% of the maximum possible, anaerobic metabolism takes over.
   b. The reactions of glycolysis occur more rapidly than aerobic ones.

6. The **oxygen debt** occurs after strenuous exercise when the muscle is deficient in its stores of creatine phosphate and glycogen and has an excess of lactic acid.
   a. Muscle fatigue often results when energy demand is greater than energy supply.
   b. Metabolic imbalances are corrected by aerobic metabolism.
   c. The lactic acid is oxidized to compounds producing ATP, and glycogen is resynthesized.
   d. Both processes are responsible for the high oxygen consumption of the resting muscle.

7. Eighty percent of the energy used in muscle metabolism is converted to heat.

B. Muscle types are adapted for special tasks

1. These specialized fibers are adapted structurally and biochemically.

2. **Red muscle fibers** contain much **myoglobin** and are categorized as **fast twitch** or **slow twitch**.
   a. Myoglobin binds, stores, and releases oxygen and is abundant in fibers that depend on oxidative metabolism.
   b. The speeds of contraction and resistance to fatigue differ for fast-twitch and slow-twitch fibers, but both have plentiful mitochondria and are highly vascularized.
   c. The speeds of contraction correlate with the rates of ATPase activity, which is

higher in fast-twitch fibers.

d. Slow fibers are resistant to fatigue when fulfilling their usual function.

3. In contrast to red fibers, white muscle fibers are adapted for fast and powerful contractions but fatigue quickly.
   a. White fibers have high ATPase activity and rapid contraction and relaxation as a result of specializations in myosin molecules and a more aggressive Ca-pumping mechanism in the SR.
   b. White fibers have fewer mitochondria.

4. Most muscles contain a mixture of the different fiber types.

5. Patterns of innervation and usage during early development play an important role in deciding the ultimate distribution of fiber types.

C. *Smooth muscle is highly economical in its energy usage.*

1. Crossbridges cycle more slowly than those in skeletal muscle and spend more time in the attached state.

2. Less energy is used for ion pumping.

3. A significant proportion of ATP is used in regulatory phosphorylation of myosin light chains and control of Ca^{2+} concentration.

D. *The pattern of use affects the development of muscle.*

1. **Hypertrophy**, an increase in mass but not number of muscle cells, results from exercise that emphasizes isometric or slow isotonic contractions.
   a. The distribution of fiber types is relatively constant.
   b. Neither the vascular supply nor the heart increases proportionately to muscle mass.

2. Exercises emphasizing isotonic contractions produce increased vascular supply to the muscles and conditioning for the heart.

3. **Atrophy** may result from long-term immobilization, paralysis, or discontinuation from exercise.

Key Terms

acetylcholine
actin
actin-linked regulation
afterload
antagonistic pair
atrophy
calmodulin
contractility
creatine phosphate
crossbridge
fatigue
force-velocity curve
gap junction
hypertrophy
intercalated disc
isometric contraction
isotonic contraction
lactic acid
length-tension curve
motor unit
myofibril
myofilament
myoglobin
myoneuronal junction
myosin
myosin-linked regulation
oxygen debt
red fibers (fast, slow)
sarcolemma
sarcomere
sarcoplasmic reticulum
summation
t-tubules
tendon
tetanus
tonus
tropomyosin
troponin
white fibers

Testing Your Knowledge

I. Matching. Select the correct term from the Key Terms.

________ 1. connective tissue attaching muscle to bone

________ 2. term describing flexor/extensor combination

________ 3. basic unit of contraction in myofiber

________ 4. comprise thin filament

________ 5. comprise thick filament

________ 6. composed of muscle cell plasma membrane and thin layer of connective fibers

________ 7. storage sites for calcium ions

________ 8. specialized area between cardiac cells providing electrical connection

________ 9. contraction in which muscle develops force but does not shorten

________ 10. contraction in which force remains constant while muscle shortens

_________ 11. one motor neuron and one or more muscle fibers

_________ 12. carries action potential to interior cell

_________ 13. contraction resulting from summation

_________ 14. produced during oxygen debt

_________ 15. fiber with fast contraction speed, low myoglobin content, and low number of mitochondria

II. Dichotomous. Circle the correct choice from the paired terms in parentheses.

1. Actions for smooth muscle are initiated primarily by the (ANS/CNS), while those for skeletal muscle are initiated by the (ANS/CNS).

2. (Multiunit/Unitary) smooth muscle is found in the small intestine.

3. (Myosin/Actin) is found in thick myofilaments and in the (A-band/I-band).

4. When a sarcomere shortens, the size of the (A-band/I-band) decreases, but the size of the (A-band/I-band) doesn't change.

5. If the afterload is greater than F_O, the contraction will be completely (isometric/isotonic).

6. In a relaxed muscle the concentration of Ca^{2+} is less than (10^{-2} mM/10^{-6} mM) but as a contracted muscle the concentration is greater than (10^{-2} mM/10^{-6} mM).

7. (Troponin/Tropomyosin) prevents actin from binding to myosin crossbridges.

8. (Actin-linked/Myosin-linked) regulation of muscle contraction occurs in skeletal muscle, while (actin-linked/myosin-linked) regulation occurs in smooth muscle.

9. (Phasic/Tonic) contractions occur in circular and longitudinal muscles in the intestine.

10. (Atrophy/Hypertrophy) of a muscle occurs when motor neurons to it are severed.

III. True/False. In the space to the left of the number write T for true or F for false.

_________ 1. Smooth muscle may contract in the absence of neuronal stimulation.

_________ 2. When a sarcomere contracts, myosin myofilaments move over the actin filaments.

_________ 3. ATPase is located in the crossbridges of actin filaments.

_________ 4. In an isometric contraction, force increases but the muscle does not shorten, whereas in an isotonic con traction, force is constant and the muscle shortens.

_________ 5. There is a direct relationship between velocity of shortening and the weight of the afterload.

_________ 6. Maximal power output is attained when muscle force is about 50% or one half its maximal value as shown by the symmetrical force-velocity curve.

_________ 7. When MLCK binds to calcium ions, it is activated to dephosphorylate myosin heads.

_________ 8. A muscle twitch occurs more rapidly than an action potential.

_________ 9. In the presence of botulinum toxin, ACh is not hydrolyzed by acetyl cholinesterase, and this causes the muscle to relax.

_________ 10. The oxygen debt refers to the lack of oxygen during strenuous exercise.

IV. Sentence Completion. Write your response in the space provided.

1. Your instructor removes a piece of muscle from a mouse and places it in physiological saline. The muscle continues to contract. Even though you don't know the source, you know it is not a _____________ muscle.

2. Tendons are unique to _____________ muscle and are attached to _________________.

3. The ______________ is the contractible unit in ______________ muscle.

4. Each myosin filament can form crossbridges with ______________ actin filaments.

5. An action potential is conducted into the depths of the muscle fiber along the ________, which lies close to the _________.

6. Each cycle of attachment-power stroke-detachment requires ___________ molecules of ATP.

7. Ca^{2+} binds to ___________ in myosin-linked regulation and to __________ in actin-linked regulation.

8. A stronger, more long-lasting contraction, called a __________, may be produced by ______________.

9. Functional syncytia are found in _________ and ___________ muscles but not _________ muscle.

10. During strenuous exercise, ATP is generated by ____________ metabolism.

V. Multiple Choice. Circle the letter of the best response.

1. Cardiac muscle may be classified as:
   a. smooth and involuntary
   b. striated and voluntary
   c. smooth and voluntary
   d. striated and involuntary
   e. visceral and involuntary

2. You are taking a laboratory practical on muscles. You must identify the type of muscle on the electron micrograph to pass the course. You see many mitochondria, striations, dyads, SR cisternae, glycogen granules, gap junctions, and branching cells. You identify the muscle as:
   a. skeletal
   b. cardiac
   c. visceral
   d. smooth

3. A common feature that is found in all three types of muscle is:
   a. multinucleate cells
   b. gap junctions
   c. intercalated discs
   d. t-tubule system
   e. myosin

4. The amount of force that a muscle can develop is:
   a. greatest when the myosin filaments are next to the Z-lines
   b. greatest when the H-band is at its widest
   c. directly related to myofilament overlap
   d. directly related to the size of the sarcomere
   e. indirectly related to the length of myosin tails

5. Optimal power output is achieved when:
   a. maximum force of the muscle is attained
   b. V_{max} is attained
   c. force is about 30% of its maximal value
   d. overlap of myofilaments is maximal
   e. an isometric contraction occurs

6. Smooth muscle uses a different type of regulation of contraction because it is lacking:
   a. tropomyosin
   b. troponin
   c. myosin
   d. actin
   e. calmodulin

7. All the following events occur in the contraction of skeletal and smooth muscles except:
   a. sliding of actin filaments over myosin filaments
   b. binding of Ca^{2+} to troponin
   c. crossbridge attachment, rotation, detachment
   d. activation of myosin
   e. ATP hydrolysis

8. Changes in contractility of cardiac muscle may occur as a result of:
   a. increasing the duration of action potentials
   b. increasing the frequency of action potentials
   c. increasing the entry of Ca^{2+} with each action potential
   d. presence of inotropic agents
   e. all of the above

9. A deficiency of calcium in the body would affect directly or indirectly all of the following events except:
   a. release of acetylcholine

b. binding of actin to myosin
c. activation of myosin in smooth muscle
d. reuptake of choline into motor neuron
e. removal of tropomyosin from myosin binding site on actin

10. White fibers and fast-twitch fibers differ in all the following features except:
    a. SR Ca^{2+} pumping
    b. source of ATP
    c. number of mitochondria
    d. myoglobin content
    e. number of capillaries

VI. Identify the structures in the diagram and match with their definitions.

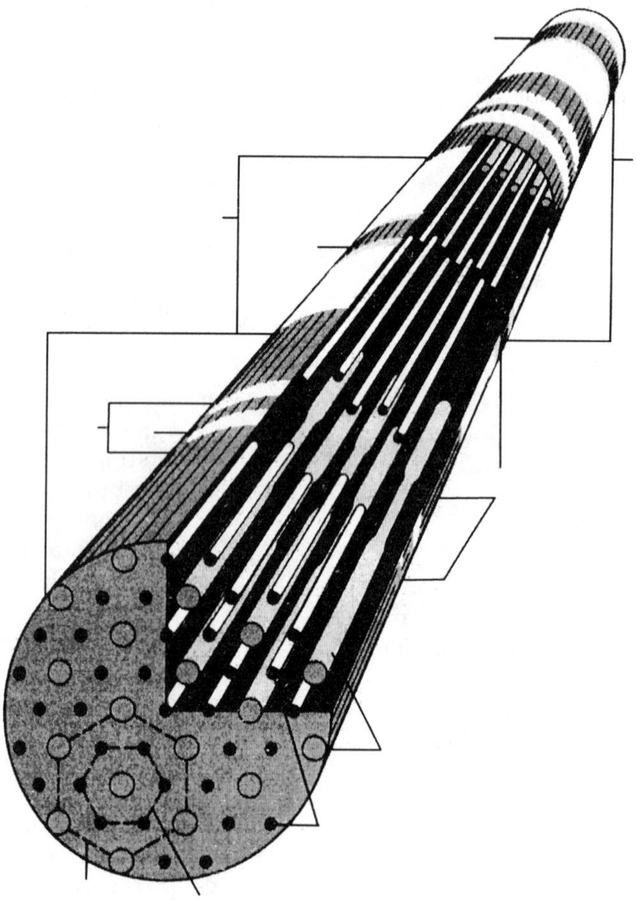

	Letter	Structure
1. composed of myosin	____	________
2. composed of actin	____	________
3. location of tropomyosin	____	________
4. site of z band	____	________
5. decreases in size in contracted sarcomere	____	________
6. stays the same during contraction	____	________

VII. For the given conditions, indicate the effect (promote, inhibit, or no effect) on contraction of a skeletal muscle.

____________ 1. deficiency of ADP

____________ 2. high quantity of lactic acid

____________ 3. nonfunctional calcium pump in SR membrane

____________ 4. deficiency of acetylcholinesterase

____________ 5. no t-system

Chapter 17

Functions of the Blood

Chapter Learning Objectives

1. Follow the development, life cycle, and destruction of red blood cells.
2. Explain the molecular structure of hemoglobin.
3. Explain the red cell indices and describe how they would vary in the following disease states: hemolytic anemia, pernicious anemia, aplastic anemia, and polycythemia.
4. Describe the different types of white blood cells, how to distinguish them, and their different functions.
5. Indicate how platelets are formed and how they function, including the actions of physiological and therapeutic anticoagulants.
6. List the plasma proteins and give the physiological functions of each.
7. List the normal constituents of blood.
8. Explain the ABO and Rh blood types and their roles in transfusion.

Expanded Chapter Outline

I. General Features of Blood

A. Blood transports molecules and heat.

1. Blood is a connective tissue composed by volume of about 45% cellular elements and 55% intercellular fluid.

2. Blood is the primary means by which substances are moved throughout the body.

3. Nutrients, electrolytes, H_2O, oxygen, CO_2, cellular waste, hormones, and chemical messengers are transported.

B. Blood cells defend against foreign agents.

1. Phagocytic blood cells remove foreign, pathogenic substances and microorganisms.

2. Other blood cells release chemicals that nullify harmful effects of foreign agents and that regulate blood flow and clotting.

C. Blood proteins help maintain normal extracellular fluid pH and osmolarity.

1. Chemical buffers in the blood regulate pH and minimize large shifts in pH during daily metabolism.

2. Blood also transports acidic and basic substances to organs of excretion.

3. Plasma proteins create an osmotic pressure that influences the movement of other solutes and water between blood and interstitial fluids and, in turn, the osmolarity of extracellular fluids.

II. Properties of Whole Blood

A. Whole blood consists of cells, platelets, and plasma.

1. Whole blood contains **erythrocytes, leukocytes,** and platelets suspended in **plasma.**

2. A variety of proteins is found in the plasma fraction.

3. Blood volume varies with age, sex, and changes in body water balances.

4. Blood distribution varies between the heart and the different types of vessels.

B. The specific gravity and viscosity of blood are determined by the cells and solute.

1. The specific gravity is usually greater than 1.0 and depends on the number of cells and concentration of chemicals dissolved in the plasma.

2. Blood **viscosity** is much greater than that of water and is determined by the number of

cells present and mutual attraction of molecules in the plasma.

C. Erythrocyte sedimentation rate is a convenient indicator of the level of inflammatory processes.

1. Sedimentation of erythrocytes proceeds in three phases: (1) formation of rouleaux, (2) rapid settling, and (3) final packing of red cell mass.

2. **Sedimentation rate** is affected by an increase in the concentration of plasma proteins, the presence of acute and chronic infections, allergy, and hyperglycemia.

*D. The **hematocrit** is the percentage of the blood volume made up of erythrocytes.*

1. Normal values vary with age, sex, body tissue, and type of blood taken (i.e., venal or arterial).

2. Abnormal hematocrit values occur in diseases, altering the number of circulating erythrocytes, the volume of plasma, or both.

III. Plasma

A. The chemical composition of plasma is normally variable.

1. Plasma is whole blood minus the formed elements. Serum is plasma minus the coagulation factors.
   a. Plasma contains water (93%) and solutes (7%), which include proteins, eletrolytes, nutrients, dissolved gases, hormones, enzymes, vitamins, pigments, minerals, and waste (see Table17-1).
   b. The composition varies as substances are exchanged between the plasma and interstitial fluid.

2. The three major fluid compartments are the blood plasma, interstitial fluid, and intracellular fluid.
   a. An osmotic equilibrium exists among the three compartments.
   b. Fluids are rapidly exchanged among the compartments.

B. Plasma proteins supply cell nutrition, serve as carriers and buffers, help clotting, and govern water distribution.

1. Plasma proteins can be divided into the following fractions: albumin, globulin, and fibrinogen, which are classified as colloids.

2. Plasma proteins play roles as:
   a. Reserve supply of amino acids for cell nutrition.
   b. Carriers for other molecules such as iron, calcium, hormones, and lipids.
   c. Buffers to maintain stable pH.
   d. Enzymes involved in the clotting process.

3. Homeostasis depends on the osmotic pressure of the plasma.
   a. Plasma proteins produce **colloid osmotic pressure.**
   b. Other solutes (e.g., electrolytes, glucose) contribute to osmotic pressure in plasma and interstitial fluid.

C. Functions of specific plasma proteins:

1. **Albumin** (60% of plasma protein) accounts for 80% of the colloid osmotic pressure.
   a. It plays an important role in transport of substances, such as pigments (bilirubin, urobilin), hormones (thyroxine), fatty acids, and bile acid salts.
   b. It helps substances (e.g., barbiturates, penicillin) dissolve in the plasma by binding to them, thereby facilitating transport.

2. **Globulins** (about 40% of total plasma) have different functions.
   a. **Alpha-1 globulins** transport thyroxine (a thyroid hormone), cortisol (adrenal cortex hormone), and vitamin B12.
   b. **Alpha-2 globulins** include **haptoglobin** (a protein preventing excretion of free hemoglobin), **ceruloplasmin** (an oxidative enzyme), **prothrombin** (a blood-clotting protein), **erythropoietin** (hormone stimulating red blood cell production), and **angiotensinogen** (a hormone precursor involved in regulating blood pressure).
   c. **Beta globulins** include most of the carriers for lipids (e.g., low-density lipoprotein), phospholipids, glycerides, lipid-soluble vitamins, and metals (e.g., copper, iron).
   d. **Gamma globulin** fraction contains the five major classes of immunoglobulins (IgG, IgM, IgD, IgA, and IgE), components of the body's immune system.

3. Fibrinogen is manufactured in the liver and is converted into fibrin during blood clotting.

IV. Hematopoiesis: The Formation of Blood Cells

A. *All blood cells and structures originate from* ***primitive pluripotential hematopoietic stem cells (PHSC).***

B. *Uncommitted stem cells, with the exception of lymphoid progenitor cells, become committed progenitor cells called* ***colony forming units*** *(CFU).*

1. CFUs may be multipotential, e.g., myeloid progenitor cells.

2. CFUs may be bipotential, e.g., CFU-GM.

3. CFUs may be unipotential, e.g., CFU-E, CFU-Me.

V. Erythrocytes: Red Blood Cells

A. *The erythrocyte shape is optimal for its function of gas transport.*

1. Their biconcave-disc shape provides a large surface area for diffusion of gases.

2. They lack cellular organelles and have a meshwork of proteins forming a cytoskeleton.

B. *The oxygen carrying capacity of blood depends on hemoglobin concentration.*

1. **Hemoglobin** is composed of a globin molecule and four attached heme groups, each of which carries an oxygen.

2. Hemoglobin readily associates and dissociates with oxygen to form oxyhemoglobin and carbon dioxide.

3. Hemoglobin also plays a role as a buffer.

C. *Not all hemoglobin molecules are alike.*

1. The **heme** part of all hemoglobins is the same, but the globin part varies depending on the composition of its subunit peptide chains.

2. Hemoglobin (Hb) varies with developmental stages.
   a. Type HbA1 (two alpha and two beta chains) is the normal adult form.
   b. Type HbF (two alpha and two gamma chains), a fetal form, has a greater affinity for oxygen than type HbA1.

3. Normal and abnormal forms of hemoglobin are genetically determined.

D. *Erythropoiesis is the formation of red blood cells.*

1. **Erythropoiesis** involves a series of steps.
   a. An uncommitted stem cell differentiates into a reticulocyte, which matures into an erythrocyte.
   b. The cell loses its nucleus and acquires hemoglobin.

E. *Iron and other substances are required for erythropoiesis.*

1. Erythropoiesis requires amino acids, copper, vitamins B_2 and B_{12}, pyridoxine, and folic acid.
   a. A stable supply of iron is maintained through a feedback mechanism involving transferrin.
   b. A deficiency of any requirement may lead to anemia.

2. Erythropoiesis is hormonally controlled by erythropoietin, which is released by the kidney under conditions of reduced oxygen tension.

F. *Aged erythrocytes are destroyed by macrophages.*

1. Erythrocytes are destroyed after about four months.

2. Important components of hemoglobin, such as heme, are recycled within the body.

G. *Red blood cell indices are used to assess anemia.*

1. **Anemia** is characterized by an abnormally low oxygen-carrying capacity of the blood.

2. It may result from a low number of erythrocytes or hemoglobin (or both), a failure of erythrocyte production, or an excessive rate of destruction.

3. Red cell indices, such as **mean corpuscular volume (MCV)** and **mean corpuscular hemoglobin concentration (MCHC)**, are used to detect abnormalities in erythrocyte size, shape, and color.

4. Iron deficiency anemia is caused by a lack of dietary iron, inadequate iron absorption, loss of iron stores, or abnormalities in iron utilization.

5. In hemolytic anemia the rate of red blood cell destruction exceeds the replacement rate.

H. Polycythemia is an increase in red blood cell count.

1. **Polycythemia** is a relative or absolute increase in the number of circulating red cells above 6 million/mm3 and results in an increase in the viscosity of blood.

2. Secondary polycythemia occurs in response to arterial hypoxia.

3. Polycythemia vera is not associated with lack of oxygen.

VI. Blood Types

A. Blood types are determined by molecules embedded in the surface of the red blood cell.

1. Incompatibilities of blood are due to the presence of agglutinogens within red blood cell membranes and specific antibodies called agglutinins in the plasma.

2. The type of **agglutinogen** or erythrocyte **antigen** is genetically determined.

3. The antigens provide the basis for classification of blood types.

B. The ABO blood group contains four major blood types.

1. A, B, AB, and O Types:

Blood Type	RBC surface antigen	Plasma antibodies
A	A antigens	anti-B
B	B antigens	anti-A
AB	A and B antigens	no antibodies to A or B
O	neither A nor B antigens	anti-A and anti-B antibodies

2. Major agglutination occurs when the recipient's antibodies complex with antigens on the surface of erythrocytes if the blood types do not match.

3. Cross-matching of blood types reduces the chances for a transfusion reaction.

C. Rh blood group contains two major blood types as determined by the presence or absence of erythrocyte antigen D

1. Antigen D is present on the erythrocyte membrane in Rh^+ blood type and absent in Rh^- blood type.

2. Incompatibility may occur between a RH^- mother and her Rh^+ fetus.
   a. An Rh^- person can produce antibodies against the Rh antigen.
   b. Rh^- blood can be transfused into an Rh^+ person, but the reverse is not possible.
   c. Erythroblastosis fetalis results when an Rh^- mother's antibodies against the Rh antigen cross the placenta and complex with the Rh antigens in the Rh^+ fetus.
   d. The chance of developing this condition is reduced by the administration of Rh antibodies (RhoGAM) to the Rh^- mother within 72 hours of giving birth to an Rh^+ child.
   e. The Rh^- mother does not build up antibodies to the Rh antigen.

VII. Leukocytes: White Blood Cells

A. Leukocytes are classified into two main categories: granular and agranular.

1. They are classified according to the shapes of their nuclei and the presence or absence of cytoplasmic granules.
   a. **Granular leukocytes** include basophils, neutrophils, and eosinophils.
   b. **Agranular leukocytes** include **monocytes** and **lymphocytes.**
   c. A **differential white blood cell count** is used diagnostically.

2. There are three types of granulocytes.
   a. **Neutrophils** actively phagocytize microorganisms, particulate matter, and fibrin. They are the first cells to arrive at the site of a bacterial invasion.
   b. **Eosinophils** detoxify foreign proteins and other substances. They are increased in numbers in allergic and autoimmune reactions and parasitic infections.
   c. **Basophils** release histamine in areas of

tissue damage to increase blood flow and attract leukocytes to the damaged area.

3. Agranulocytic **T-lymphocytes** have important functions in the immune system.
   a. T-lymphocytes originate from stem cells in the bone marrow but complete their development in the thymus, where they acquire receptors enabling them to recognize antigens.
   b. On the presence of a foreign antigen preconditional T-lymphocytes may (1) attach directly; (2) release lymphokines or toxins that neutralize the antigen or activate complement; (3) stimulate B-lymphocytes; or (4) produce a clone of lymphocytes responsive to the antigen, i.e., confer cell-mediated immunity.

4. **B-lymphocytes** originate in the bone marrow and later migrate to lymphoid tissues in the spleen, tonsils, lymph nodes, and walls of the small intestine.
   a. When stimulated by a foreign antigen, B-lymphocytes become specialized to produce antibodies against the antigen.
   b. They also become memory cells.

5. **Monocytes** originate in bone marrow and develop into **macrophages**, large phagocytic cells.

B. ***Leukopoiesis** is the generation of white blood cells.*

1. Granulocytes and monocytes mature in the marrow, but T- and B-lymphocytes mature in the thymus and lymphoid tissues, respectively.

2. **Granulopoiesis** is stimulated by leukocyte-inducing factor (LIF).

3. The life span of a leukocyte depends on the role it plays in defense. Neutrophils are short-lived because of their phagocytic activities.

C. ***Leukocytosis** is an increase in white blood cell count.*

1. It is an abnormal increase in the total number of circulating leukocytes and may have physiological or pathological causes.

2. **Leukemia** is characterized by the prolifera tion of immature and functionally impaired hematopoietic cells.

D. ***Leukopenia** is a reduction in white blood cell count.*

1. It may result from bone marrow defects.

2. It may also result from arrest of cell development and maturation or excessive destruction.

VIII. Thrombocytes (Platelets)

A. *Platelets are circulating fragments of megakaryocytes that remain in the marrow.*

1. They are anucleate and packed with granules, vesicles, microfilaments, microtubules, and occasionally mitochondria.

2. Platelets play a role in maintaining hemostasis.
   a. They release chemicals, which stimulate contraction of injured blood vessels.
   b. They release platelet-derived growth factor, which promotes repair of injured tissue.

IX. Hemostasis: Mechanisms to Prevent Blood Loss

A. *Four interrelated events constitute hemostasis: (1) local vasoconstriction, (2) formation of a platelet clump, (3) formation of a blood clot, and (4) clot retraction and dissolution.*

1. Activated platelets produce thromboxane A^2, a vasoconstrictor and platelet aggregator.
   a. Vasoconstriction is an autonomic nervous system response that results from local muscle spasm in the wall of the injured blood vessels.
   b. It is maintained by vasoconstrictor chemicals released from platelets accumulating at the damaged site.

2. At the damaged site platelets release stored chemicals and form an aggregate or plug.

B. *Blood clot production occurs in three stages.*

1. Stage I is the formation of a **prothrombin converting factor** and occurs in a sequence of steps in the extrinsic pathway:

   a. Contact of blood with the injured tissue.
   b. Release of tissue lipoprotein **thromboplastin (factor III)**.
   c. Interaction of factor III with factor VII and calcium ions to form factor X (Stuart factor).
   d. Factor X plus Ca^{2+} complexes with factor V to form prothrombin converting factor.

2. The sequence of steps in intrinsic coagulation is:
   a. Activation of factor XII by collagen, fibrin, or platelet membrane during platelet aggregation, stress, anxiety, and fear.
   b. Factor XII (Hageman factor) activates factor XI, PTA.
   c. Factor XI plus Ca^{2+} activates factor IX, which interacts with factor VIII in the presence of Ca^{2+} to form a complex activating factor X.

3. Stage II is the conversion of **prothrombin** to **thrombin**.
   a. This conversion to thrombin, an enzyme, occurs in the presence of Ca^{2+} and prothrombin-converting factor.
   b. Thrombin increases its own rate of production by converting factor V into accelerin.

4. Stage III is the conversion of **fibrinogen** to **fibrin**.
   a. Thrombin converts fibrinogen to fibrin monomers, which spontaneously polymerize to form the clot.
   b. Thrombin activates fibrin-stabilizing factor, which in the presence of Ca^{2+} stabilizes the fibrin polymer.

5. Clot retraction is caused by platelet factor, thrombosthenin, which requires the presence of thrombin and ATP.

6. **Fibrinolysis**, the dissolution of a clot, may occur enzymatically by plasmin or by neutrophils and macrophages.
   a. **Plasmin** formation requires plasminogen activator.
   b. Enzymes to convert the precursor to activator are found in some bacteria, urine, tears, and saliva.

C. Anticoagulants inhibit blood clotting.

1. Antithromboplastin, antithrombin, and heparin are physiological anticoagulants.

2. Therapeutic coagulants are classified on the basis of action into three groups: (a) agents that depress the action of procoagulants; (b) agents that remove calcium from the blood; and, (c) agents that inhibit synthesis of prothrombin in the liver.

3. Abnormalities of hemostasis occur as a result of: (a) platelet defects **(thrombocytopenia)**; (b) coagulation defects **(hemophilia)**; (c) excessive production of anticoagulants (hemorrhagic disease); or (d) a combination of these.

Key Terms

agglutination
agranulocyte
albumin
anemia
antibody
anticoagulant
antigen
atherosclerosis
B-lymphocyte
basophil
clot leukocytosis
coagulation
colony-forming unit
eosinophil
erythrocyte
erythropoiesis
erythropoietin
fibrin
fibrinogen
fibrinolysis
granulocytes
granulopoiesis
hematocrit
hematopoiesis
hemolysis
hemophilia
hemostasis
interleukin
leukemia
leukocyte
leukopenia
leukopoiesis
lymphokines
macrophage
mean corpuscular hemoglobin (MCH)
mean corpuscular hemoglobin concentration (MCHC)
mean corpuscular volume index (MCV)
megakaryocyte
monocyte
neutrophil
oxygen carrying capacity
phagocytosis
plasma
platelet (thrombocyte)
pluripotent hemopoietic stem cell
polycythemia
progenitor cell
reticulocyte
Rh factor
serum
T-lymphocyte
thrombin
thrombocytopenia
thromboplastin
thrombus
transferrin
viscosity

Testing Your Knowledge

I. Matching. Select the correct term from the Key Terms.

_________ 1. red blood cells

_________ 2. development of red blood cells from stem cells

_________ 3. measure of blood's resistance to flow

_________ 4. oxygen carrying molecule

_________ 5. term defining destruction of erythrocytes

_________ 6. blood condition characterized by low oxygen carrying capacity

_________ 7. most numerous type of granulocyte

_________ 8. granulocytes that increase in number in allergic and autoimmune reactions

_________ 9. deficiency of platelets

_________ 10. concentration of hemoglobin in erythrocytes

_________ 11. disease characterized by proliferation of hematopoietic cells that are immature

_________ 12. anucleate, membrane bodies that play a role in clot formation

_________ 13. converts fibrinogen to fibrin

_________ 14. carrier of iron in blood

_________ 15. major plasma protein

II. Dichotomous. Circle the correct choice from the paired terms in parentheses.

1. (Hypervolemia/Hypovolemia) describes a blood volume above normal.
2. (Erythrocytes/Leukocytes) are found in the buffy coat of sedimented blood.
3. (Serum/Plasma) is the liquid part of the blood and lacks coagulation factors.
4. Erythrocytes placed in a solution of 0.7% NaCl would (crenate/ hemolyze).
5. The (alpha/beta) chains differ between hemoglobin A and hemoglobin S.
6. Normocytic, normochrome erythrocytes would be seen in a person with (iron deficiency/aplastic) anemia.
7. Secondary polycythemia occurs in response to (low/high) oxygen.
8. When you perform a differential white blood cell count on a healthy person, you expect more than 50% of cells to be (lymphocytes/neutrophils).
9. (Leukopenia/Leukopoiesis/Leukocytosis) is an abnormal increase in the total number of circulating leukocytes.
10. (Neutrophils/Platelets) are the first cells to arrive at the site of an injured blood vessel.
11. (Hemophilia/Anemia) results from a lack of clotting factors VIII and IV.

III. True/False. In the space to the left of the number write T for true or F for false.

_______ 1. An elevated hematocrit would be an indicator for anemia.

_______ 2. Water, electrolyte, and protein concentrations of plasma and interstitial fluids are nearly identical.

_______ 3. The erythrocyte has a high ratio of surface area to volume.

_______ 4. Hemoglobin F has a greater affinity for oxygen than hemoglobin A.

_______ 5. Erythropoietin is released by the kidney in response to elevated blood oxygen tension.

_______ 6. A person with blood type O could receive whole blood from a person with blood type A or B without serious medical consequences.

_______ 7. Rh^+ blood may be given to Rh^+ or

Rh^- recipients without medical consequences.

8. Platelets give rise to megakaryocytes, which play a role in clotting.

9. Leukemia is a condition characterized by the proliferation of immature leukocytes.

10. During coagulation fibrin is converted to fibrinogen by thrombin.

IV. Sentence Completion. Write your response in the space provided.

1. A person with arthritis would most likely have a(n) __________ sedimentation rate.

2. Normal plasma is about ____________ % water and _________________ % solutes.

3. The O_2 carrying capacity of normal adult blood is _________________ ml O_2/dL.

4. Pernicious anemia may result from a deficiency of __________ or ____________ .

5. Agglutination would occur between erythrocytes from type A blood and serum from type _______________ blood.

6. ______________ are cells whose granules contain histamine and heparin.

7. ______________ are conditioned in the thymus and release chemicals called ______________ .

8. B and T lymphocytes can be distinguished by their ______________ .

9. Vitamin ____________ and ____________ ions are essential for hemostasis.

10. In stage II coagulation ____________ , a plasma globulin made in the liver and known as factor II, is converted to _______ , which acts to convert _______ to _______ .

11. ___________ converts ___________ to ___________ and is responsible for most of physiological fibrinolysis.

12. ___________ is a condition in which the ___________ count falls below 50,000 cells/mm^3.

V. Multiple Choice. Circle the letter of the best response.

1. The specific gravity of blood ranges between:
   a. 1.05 and 1.06
   b. 7.4 and 7.5
   c. 4.5 and 5.5
   d. 6 and 8
   e. 1.9 and 2.0

2. Which of the measures of blood decreases with an increase in erythrocyte count?
   a. specific gravity
   b. viscosity
   c. sedimentation rate
   d. hematocrit
   e. none of the above

3. Under a condition of protein starvation, the osmotic pressure of interstitial fluid is ________ that of the plasma.
   a. same as
   b. greater than
   c. less than

4. The A/G ratio of a person with mononucleosis would be __________ that of a healthy individual.
   a. greater than
   b. less than
   c. equal to
   d. twice as great as
   e. the inverse of

5. A deficiency of transferrin would have the greatest effect on:
   a. hematopoiesis
   b. erythropoiesis
   c. leukopoiesis
   d. lymphopoiesis
   e. megakaryocytopoiesis

6. You are viewing a smear of human blood using a microscope. You notice a larger than expected number of reticulocytes and erythrocytes. For this person, which of the following molecules would you expect to be elevated above normal?
   a. colony-stimulating factor
   b. interleukin
   c. erythropoietin
   d. thrombopoietin
   e. fibrinogen

7. Consider a person whose liver has been damaged by alcohol. Which of the following

abnormal laboratory conditions would you expect?
a. hyperbilirubinemia
b. decreased albumin
c. decreased fibrinogen
d. decreased iron
e. all of the above

8. Cytoplasmic granules containing digestive enzymes and antibacterial proteins are found in
a. lymphocytes
b. monocytes
c. neutrophils
d. eosinophils
e. basophils

9. Information about which one of the following would help you to distinguish among the five different types of leukocytes?
a. presence and content of granules
b. place of origin in the body
c. shape of nucleus
d. ability to perform phagocytosis
e. motility

10. The correct sequence of events in hemostasis is:
a. vasoconstriction → coagulation → aggregation → clot formation → clot retraction
b. aggregation → coagulation → vasoconstriction → clot dissolution → clot retraction
c. vasoconstriction → aggregation → coagulation → clot retraction → clot dissolution
d. coagulation → aggregation → vasoconstriction → clot formation
e. none of the above

11. In the absence of factor VIII, which of the following reactions will not occur?
a. activation of Hageman factor
b. formation of plasma thromboplastin antecedent
c. formation of Christmas factor
d. formation of prothrombin converting factor
e. formation of thrombin

12. If the hematocrit is 42% and RBC count is $5.4 \times 10^6/\mu l$, the mean cell volume (MCV) is:
a. 42 μm^3
b. 54 μm^3
c. 78 μm^3
d. 87 μm^3
e. none of the above

13. If the hemoglobin is 15 g/dL and RBC count is $5.4 \times 10^6/\mu l$, the MCH is:
a. 7 μμg.
b. 14 μμg.
c. 21 μμg.
d. 28 μμg.
e. 35 μμg.

14. If the hemoglobin is 15 g/dL and the hema tocrit is 42%, the MCHC is:
a. 15%
b. 25%
c. 36%
d. 42%
e. none of the above

VI. Complete the Table.

Type of Cell	Place of Origin	Presence of Granules	Function	Disease Association
erythrocyte				
neutrophil				
basophil				
eosinophil				
T-lymphocyte				
B-lymphocyte				
thrombocyte				
monocyte				

Chapter 18

The Heart

Chapter Learning Objectives

1. Characterize the circulatory system as two pumps with associated vessels.
2. Follow a red blood cell through a complete circuit of the body.
3. Describe how blood pressure is generated and how it controls movement of blood through the body; describe the relationship between blood flow, driving pressure, and resistance.
4. Diagram the heart and label the significant structures.
5. Explain how an action potential is generated in a heart muscle and give the role of the nodes in generating and transmitting action potentials.
6. Diagram and explain an electrocardiogram.
7. Describe the sequence of events in a cardiac cycle and explain the regulation of both the cardiac cycle and the cardiac output.
8. Indicate how heart rate and stroke volume are regulated.

Expanded Chapter Outline

I. Overview of the Cardiovascular System

A. *The systemic circulation carries oxygenated blood to the tissues of the body.*

1. The vessels of the arterial system carry blood from the heart.
2. The vessels of the venous system carry blood back to the heart.
3. The pathway to blood flow through the **systemic circulation** is:

> aorta → arteries → arterioles → capillaries → venules → veins → inferior and superior vena cavae

B. *The pulmonary circulation carries unoxygenated blood through the lungs for gas exchange.*

1. The pathway of blood flow through the **pulmonary circulation** is:

> pulmonary trunk → pulmonary arteries → capillaries → pulmonary veins

2. The pulmonary capillaries are the sites of gas exchange between blood and air in the lungs.

C. *Blood flow is determined by blood pressure and the resistance to blood flow.*

1. Blood flow occurs in response to the pres sure gradient or difference between two points and is proportional to the difference in blood pressure between the two points.
2. **Resistance** measures opposition to blood flow.
3. The relationship between blood flow (F), driving pressure (P_1–P_2) and resistance (R) is:

$$F = P_1 - P_2 / R$$

II. The Heart: A Muscular Pump

A. *The heart consists of two separate pumps that maintain the unidirectional flow of blood.*

1. The right and left **atria** receive blood from the systemic and pulmonary circulations, respectively.
2. The right and left **ventricles** pump blood into the pulmonary and systemic circulations, respectively.
3. The unidirectional flow of blood through the heart is:

right atrium → tricuspid valve → right ventricle → pulmonary valve → pulmonary circulation → left atrium → mitral valve → left ventricle → aortic valve → systemic circulation

4. The response of a **valve** depends on the direction of blood pressure gradient across the valve to ensure unidirectional blood flow.

5. The walls of the heart are composed of three distinct layers:
   a. **Endocardium** (the inner lining)
   b. **Myocardium** (cardiac muscle) whose thickness varies among the chambers
   c. **Epicardium** (outer surface)

B. *The sequential contraction of the atria and ventricles is activated by a wave of electrical depolarization that is initiated by pacemaker cells in the sinoatrial node in the right atrium.*

1. The action potential can be broken down into the following phases, depending on changes occurring in the membrane:
   a. Phase zero is the rapid depolarization caused by influx of sodium ions.
   b. Phase 1 is the repolarization when sodium permeability declines.
   c. Phase 2 is a plateau with an influx of Ca^{2+} and efflux of K^{+} .
   d. Phase 3 is a repolarization to original membrane potential.
   e. Phase 4 is the steady membrane potential (resting potential) at original value until next beat.

2. The phases are associated with absolute and relative refractory periods.
   a. The **relative refractory period** is followed by a period of supernormal excitability.
   b. During the **absolute refractory period,** the muscle cannot be stimulated.

3. Cardiac muscle cells communicate directly with one another as a functional syncytium.
   a. The heart is stimulated at one point and the message is relayed to the rest of the heart muscle.
   b. The heart has no motor nerve supply.

4. The action potential is initiated by pacemaker cells located in the sinoatrial node (SA) in the right atrium.
   a. There are several pacemaker cells, but only one is active at one time.
   b. A pacemaker cell undergoes an action potential, repolarization to its diastolic potential, and depolarization to its threshold.

5. The autonomic nervous system controls the rate of the SA node.
   a. Impulses from the parasympathetic system decrease the rate of diastolic depolarization and heart beat.
   b. Impulses from the sympathetic system increase diastolic depolarization and heart beat rate.

6. Heart rate is affected by chronotropic agents and inotropic agents that affect the strength of contraction.

7. The sequence of cardiac activation is:

SA node pacemaker → right atrial muscle → left atrial muscle → right atrial muscle → atrioventricular (AV) node in the right atrium → bundle of His → right and left bundle branches to each ventricle → bundle branches → array of Purkinje fibers on endocardial surfaces → muscle of ventricular wall

   a. As the wave of depolarization passes from the base to the apex, a wave of repolarization passes in the opposite direction.
   b. The electrical activity of the heart can be detected at the body surface by **electrocardiography**.

C. *Each beat of the heart involves integrated electrical, biochemical, and mechanical events that are collectively referred to as the* ***cardiac cycle.***

1. The ventricular pumping process consists of four steps:
   a. Ventricular filling
   b. Ventricular contraction
   c. Ventricular ejection
   d. Ventricular relaxation

2. **Diastole** and **systole** refer to periods of the cardiac cycle during which ventricles are filling with blood, actively contracting, and pumping blood out of the heart.
   a. The end of ventricular filling corresponds to the end of diastole.
   b. The first distinct heart sound occurs at the time of closure of the AV valves at the beginning of systole.
   c. The second distinct heart sound occurs at the time of closure of the semilunar valves and defines the end of systole.

D. The heart is regulated by changing both the strength and the rate of contractions to produce whatever cardiac output is required.

1. **Cardiac output** (CO) equals the amount of blood pumped out of the left ventricle and into the aorta in 1 minute.

2. **Stroke volume** (SV) is the volume of blood ejected in a single beat and is equal to the end-diastole volume (EDV) minus end-systole volume (ESV).

3. Cardiac output equals heart rate times stroke volume:

$$CO = HR \times SV$$

4. Heart rate is determined by the rate of depolarization of the SA node.

5. The rate of depolarization can be altered by:
   a. Stimulation of parasympathetic and sympathetic nerve fibers to the heart.
   b. Presence of hormones, such as epinephrine.
   c. Plasma electrolyte concentrations.
   d. Body temperature.

6. Stroke volume is regulated by intrinsic mechanisms and extrinsic factors, such as neural stimulation and circulating hormones.
   a. The Frank-Starling law states that the heart adjusts its output in response to changes in input and describes how changes in end-diastolic volume result in changes in stroke volume.
   b. Changes in contractility result from changes in intracellular calcium and do not require a change in end-diastolic volume.

7. Cardiac muscle relies on continuous oxidative metabolism of circulating glucose and fatty acids to produce ATP.

8. The ability of the heart to contract depends on the availability of oxygen.

Key Terms

aortic valve
arrhythmia
arteries
arterioles
atrioventricular (AV) node
atrioventricular valve
atrium
blood pressure
bradycardia
bundle branches
bundle of His
cardiac cycle
cardiac output (CO)
coronary vessels
diastole, diastolic pressure
electrocardiogram (ECG)
endocardium
$F = P_1 - P_2$
Frank-Starling law
mitral valve
myocardium
P-wave
pacemaker
pressure gradient
pulmonary circulation
pulmonary valve
Purkinje fibers
QRS complex
refractory period
semilunar valve
sinoatrial node (SA node)
stroke volume (SV)
systemic circulation
systole, systolic pressure
T-wave
tachycardia
tricuspid valve
vascular resistance
veins, venules
ventricle
ventricular fibrillation

Testing Your Knowledge

I. Matching. Select the correct term from the Key Terms.

________ 1. receiving heart chamber

________ 2. pumping heart chamber

________ 3. located between atrium and ventricle

________ 4. located between right ventricle and pulmonary artery

________ 5. muscular wall of heart

________ 6. supply blood to heart muscle

________ 7. pacemaker of heart

________ 8. specialized muscle fibers that spread impulse over endocardial surface of ventricles

________ 9. recording of heart beat

________ 10. period when ventricles are filling

________ 11. period when ventricles are pumping

________ 12. period when aortic pressure is greater than left ventricular

________ 13. amount of blood ejected in a single beat

________ 14. rapid heart rate

________ 15. equals number of beats/minute times amount of blood ejected in a single beat

II. Dichotomous. Circle the correct choice from the paired terms in parentheses.

1. The pulmonary artery carries (oxygenated/unoxygenated) blood.

2. Blood flow is (directly/inversely) related to resistance for any given pressure difference.

3. When blood pressure in the atria is greater than in the ventricles, (aortic and pulmonary/tricuspid and bicuspid) valves open.

4. During the (absolute/relative) refractory period, cardiac muscle cannot be restimulated.

5. Stimulation via the sympathetic nervous system (slows down/speeds up) diastolic depolarization of pacemaker cells and (increases/decreases) heart rate.

6. During diastole the pressure in the atria is (greater/less) than the pressure in the ventricle.

7. The closure of the semilunar valves marks the end of (diastole/systole), and the closure of the atrioventricular valves marks the end of (diastole/systole).

8. Norepinephrine (increases/decreases) the rate of spontaneous depolarization of the SA nodal cells and (increases/decreases) heart rate.

9. There is a (direct/inverse) relationship between diastolic volume of the heart and stroke volume.

10. The energy for contraction of cardiac muscle comes from (aerobic/anaerobic) metabolism of glucose.

III. True/False. In the space to the left of the number write T for true or F for false.

______ 1. Blood flow through the liver is always equal to blood flow through the intestine.

______ 2. If blood pressure in the right atrium is higher than pressure in the right ventricle, the pulmonary valve opens.

______ 3. Rapid depolarization of cardiac muscle in phase zero is due to increased membrane permeability to potassium ions.

______ 4. The rate of the heart is regulated by the rate at which the diastolic potential reaches the threshold.

______ 5. The Purkinje fibers and bundle branch fibers are specialized neurons.

______ 6. Atrial contraction increases ventricular filling by 80%.

______ 7. The cardiac output of the left ventricle is greater than that for the right ventricle.

______ 8. From the Frank-Starling law, if diastolic volume of the heart increases, the stroke volume increases.

______ 9. Increased vigor of contraction is always associated with increased end-diastolic volume.

______10. An increase in contractility is associated with a decreased end-systolic volume.

IV. Sentence Completion. Write your response in the space provided.

1. The ____________ receives blood from the vena cavae, and the ____________ receives blood from the pulmonary vein.

2. After an action potential is generated, the pacemaker cell is repolarized to a value

called the ____________.

3. Pacemaker cells are located in the ____________ and ____________.

4. During ventricular filling, a ____________ pressure gradient exists across the semilunar valves such that ____________ pressure is greater than ____________ pressure.

5. Systole occurs in the ____________ of the ECG.

6. The ____________ is the volume of blood ejected in a single beat and is equal to EDV minus ____________.

7. Cardiac output equals ____________ times ____________.

8. Acetylcholine ____________ pacemaker cells and ____________ heart rate.

9. Changes in contractility result from changes in the ____________ concentration in the cell.

10. The effect of shortening the duration of systole is to ____________ the duration of diastole.

V. Multiple Choice. Circle the letter of the best response.

1. The heart chamber with the thickest myocardium is the:
   a. left atrium
   b. right atrium
   c. left ventricle
   d. right ventricle

2. A drug blocking the influx of Ca2+ would have its greatest effect during phase (s):
   a. zero
   b. 1
   c. 2
   d. 3
   e. 4

3. Which sequence correctly represents the succession of action potentials in one heart beat? (AV = atrioventricular node; SA = sinoatrial node)
   a. AV → Purkinje fibers → bundle branch
   b. SA → bundle branch → AV
   c. SA → AV → bundle branch
   d. SA → bundle branch → Purkinje fibers
   e. AV → SA → bundle branch

4. Which of the following sets is mismatched?
   a. ECG P-wave: atria depolarize
   b. ECG QRS complex: ventricles depolarize
   c. ECG T-wave: ventricles repolarize
   d. abnormally shaped QRS complex: defect in sinoatrial node
   e. absent QRS complex: blocked conduction between atria and ventricular muscle

5. Which of the following sets is mismatched? (AV = atrioventricular SL = semilunar)
   a. ventricular fibers; open AV valves; closed SL
   b. isovolumetric ventricular contraction; open AV valves; open SL valves
   c. ventricular ejection; closed AV valve; open SL valves
   d. isovolumetric ventricular relaxation; closed AV valves; closed SL valves
   e. atrial contraction; open AV; closed SL

6. If stroke volume is 70 ml/beat and heart rate is 70 beats/minute, cardiac output equals:
   a. 4900 ml/min
   b. 2450 ml/min
   c. 1700 ml/min
   d. 7000 ml/min
   e. none of the above

7. A more rigorous ventricular contraction may occur from:
   a. an increase in cardiac muscle fiber length
   b. hypertrophy
   c. an increase in contractile protein content
   d. a and b are correct
   e. a, b, and c are correct

8. Under which condition would the heart require the most ATP if the cardiac output were the same?
   a. increased heart rate; decreased stroke volume
   b. decreased heart rate; increased stroke volume
   c. increased blood pressure
   d. decreased blood pressure
   e. none of the above

9. You are being chased by an angry dog. Which of the following occurs?
   a. increased rate of contraction

b. increased rate of relaxation
c. decreased systolic duration
d. increased cardiac output
e. all of the above

10. Under which condition(s) would cardiac output decrease?
    a. increased stroke volume
    b. increased heart rate
    c. acetylcholine
    d. increased end-diastolic volume
    e. increased aortic pressure

11. Which pathway represents the flow of erythrocytes through the pulmonary circuit? (Note: Some structures in the circuit have been omitted.) (RA = right atrium; RV = right ventricle; LA = left atrium; LV = left ventricle)
    a. RA → RV → pulmonary vein → lung → LA → LV
    b. LA → LV → pulmonary artery → lung → RA → RV → aorta
    c. RA → pulmonary vein → lung → RV → LA → LV → aorta
    d. LA → LV → pulmonary artery → lung → RV → aorta
    e. RA → RV → pulmonary artery → lung → pulmonary vein → LA → LV

VI. Label the Diagrams.

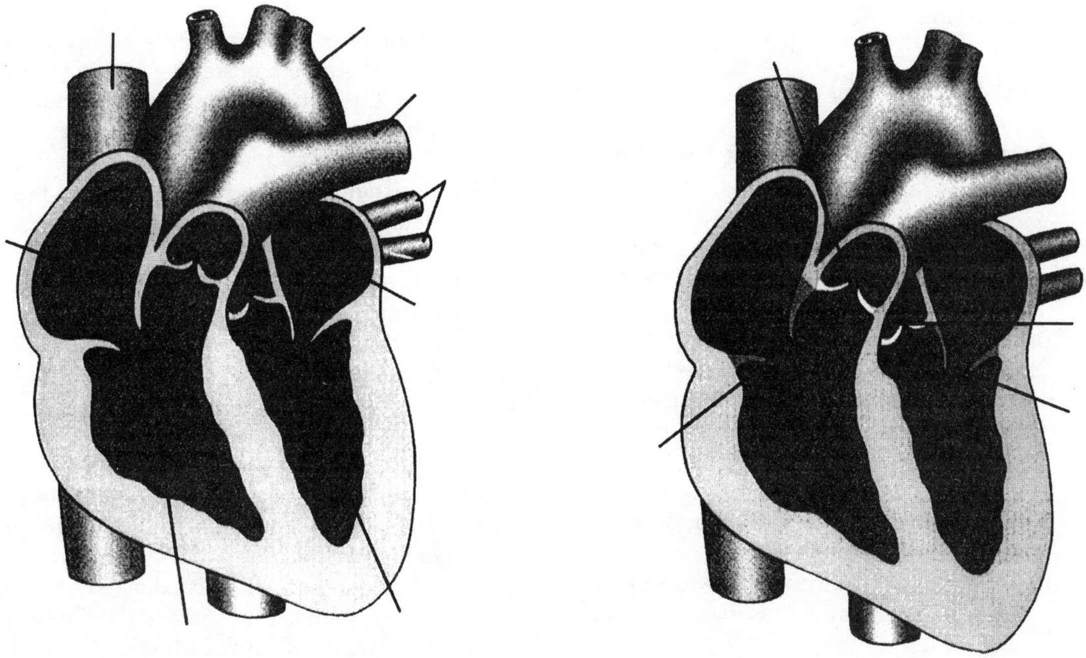

Chapter 19

The Circulation

Chapter Learning Objectives

1. Explain blood pressure and how to determine mean arterial pressure (MAP). Describe how it is affected by changes in blood flow.
2. Use the analogy of the water system to explain the regulation of blood flow.
3. Indicate how hormones and nerves control blood.
4. Explain the value and function of the capillaries and describe three different mechanisms by which materials cross the capillary wall.
5. Explain Starling's hypothesis.
6. Examine in detail the venous system.
7. Outline the basic features and functions of the lymphatic system.
8. Explain the action of baroreceptors.
9. Discuss the controls over the cardiovascular system.
10. Describe anomalies of the cardiovascular system, in particular, cardiovascular responses to changes in blood pressure, hemorrhage, hypotension, exercise, and heart failure.

Expanded Chapter Outline

I. The Arterial System: Carrying Blood to the Tissue

A. Blood pressure in the arteries propels the blood through the capillaries and back to the heart.

1. The equation for blood flow (F), $F = P_1 - P_2 / R$, can be modified to F = arterial pressure/ Resistance (R), if the pressure in the vena cava (P_2) is zero and arterial pressure (P_1) is equivalent to aortic pressure.

2. **Pulse pressure** is the difference between systolic and diastolic pressures.
   a. Energy stored in the elastic walls of the arteries during systole is dissipated during diastole.
   b. The magnitude of pulse pressure is determined by the stroke volume and the distensibility of the arteries.

3. Measurement of arterial pressure is done with a stethoscope and sphygmomanometer

4. The **mean arterial pressure (MAP)** is calculated by the equation: MAP = 1/3 pulse pressure and **diastolic pressure**.

5. Because MAP is related to blood flow, the equation for flow can be modified as:
   F = arterial pressure/R = MAP/R.

B. The major site of resistance to blood flow is the arteriole.

1. **Resistance** (R) to flow through a blood vessel is inversely proportional to the radius raised to the fourth power:

$$R = \boxed{\frac{1}{\text{radius}^4}}$$

2. Blood flow through different organs and tissues can be adjusted by altering blood flow through arterioles by contracting smooth muscle cells in their walls.

3. Resistance to blood flow through a tissue is determined almost entirely by resistance of the arterioles.

4. Overall resistance from the aorta to the heart is total peripheral resistance (TPR). Considering this,

$$F = MAP/TPR$$

5. Major points of the water system analogy that provide an insight into the regulation of blood flow are:
   a. Local blood flow is adjusted to demand by changing the diameter of the arterioles.
   b. Cardiac output is adjusted to match

demand by monitoring mean arterial pressure.
c. If demand for flow does not exceed cardiac output, individual tissue beds can function autonomously.
d. If demand exceeds flow, then extrinsic regulatory mechanisms redistribute flow to prevent pressure loss and flow.

C. Local factors such as blood pressure and metabolites autoregulate blood flow to some vascular beds.

1. In its normal resting state, the arteriolar smooth muscle is contracted. This is referred to as resting vascular tone.

2. Vascular beds can maintain a constant blood flow through an autoregulatory mechanism involving myogenic activity of arteriolar smooth muscle cells.

3. Increased blood flow to an organ or tissue through **vasodilation** can be caused by:
   a. Increased carbon dioxide or decreased oxygen.
   b. Increased adenosine.
   c. Increased hydrogen ion and potassium ion concentration.
   d. Histamine.

4. Other vasoactive substances are Prostaglandins, endothelial-derived relaxing factor (EDRF), nitric oxide, and endothelin.

D. Nerves and circulating hormones coordinate the blood flow through all of the vascular beds with the pumping activities of the heart.

1. Redistribution of blood flow requires communication between the various users, such as central nervous system, autonomic nervous system, and circulating hormones.

2. Arteriolar smooth muscle is innervated by sympathetic nerve fibers, which, when stimulated, cause **vasoconstriction**.

3. The effect of epinephrine on myogenic activity depends on the presence and number of alpha and beta receptors on smooth muscle cells.
   a. The sympathetic transmitter norepinephrine binds to alpha receptors to cause vasoconstriction.
   b. The hormone epinephrine binds to beta receptors to cause vasodilation.

4. **Atrial natriuretic peptide** causes vasodilation and increased excretion of salt and water by the kidney.

II. The Capillary System: The Exchange of Molecules

A. The exchange of molecules between the blood and the extracellular space occurs primarily in the capillary beds.

1. Most cells are within 0.02 mm of a capillary.

2. Capillary walls are one cell thick.

B. Blood flow through capillaries is controlled by the contraction of vascular smooth muscle at the level of the arteriole and the precapillary sphincter.

1. Blood flow through the capillary network is regulated by arteriolar diameter.

2. **Vasomotion** of the **precapillary sphincter** is influenced by the metabolic state of the tissue.

C. The absence of smooth muscle cells distinguishes the capillary from other small vessels.

1. The capillary wall is made up of endothelial cells and contains no smooth muscle.

2. Endothelial pores (clefts), whose size and number vary greatly among different tissues, allow for the passage of molecules.

D. Molecules exchange across the capillary wall by transcytosis, free diffusion, and bulk flow.

1. Transcytosis, an active process by which substances are shuttled in discrete vesicles or through transient vesicle-derived channels, accounts for a small portion of total exchange.

2. Molecules diffuse across capillary walls either through water-filled pores or directly through endothelial cells, depending on their relative solubility in water and lipids.

3. Several factors contribute to the rapid diffusional exchange of substances across the capillary wall:
   a. Short diffusion distances.
   b. Extensive surface area of capillaries.
   c. Slow rate of blood flow through the capillaries.

4. Exchange of water and dissolved solutes occurs through endothelial pores by bulk flow.
   a. Bulk flow results from a **hydrostatic pressure** differential across the capillary wall.
   b. **Osmotic pressure** of plasma proteins prevents loss of vascular volume.

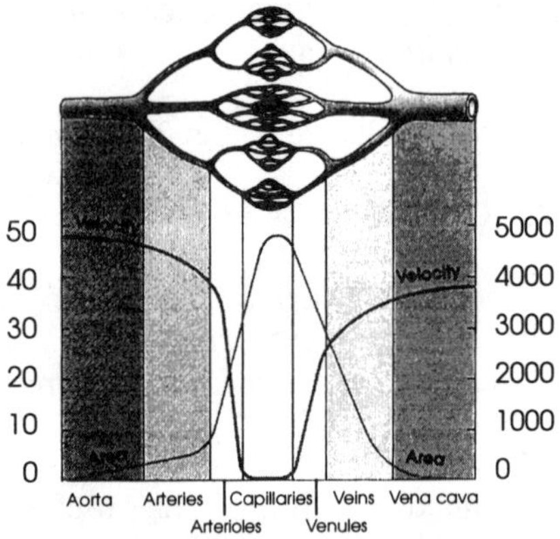

5. **Starling's hypothesis** describes the balance of hydrostatic and oncotic pressures across the capillary endothelium.
   a. Vasoconstriction promotes absorption of fluid, whereas vasodilation promotes filtration across the capillary endothelium.
   b. Net exchange across capillary endothelium is equal to filtration forces: $(\mathbf{P_c} + \pi_{if}) - (\mathbf{P_{if}} + \pi_p)$. P_c and P_{if} equals hydrostatic pressures in the capillary and interstitial fluid, respectively; π_{if} and π_p equals colloid osmotic pressure (oncotic pressure) in the interstitial fluid and plasma, respectively

III. The Venous System: Returning Blood to the Heart

A. The veins provide a conduit for the return flow of blood to the heart, and they also function as a variable capacity reservoir for blood.

1. **Venules** also contain endothelial clefts and together with capillaries are often referred to as exchange vessels.

2. Veins function as a low-pressure reservoir for blood and are often referred to as **capacitance** vessels.

3. Through contraction of their smooth muscle, veins can alter their capacity and redistribute blood.

B. The return flow of blood to the heart is facilitated by the large diameter of the veins, one-way venous valves, skeletal muscle contraction, and respiration.

1. Valves prevent back flow of blood away from the heart and in conjunction with muscular contraction contribute to the flow of blood back to the heart.

2. Venous return is assisted by one-way valves and by changes in intrathoracic pressure during breathing.

IV. The Lymphatic System

A. Lymphatic vessels begin as blind sacs that empty into lymphatic vessels.

1. The **lymphatic system** returns to the circulatory system the fluid and plasma proteins that leaked from the capillary into the interstitial space.

2. Lymphatic capillaries, which are composed of a single layer of endothelial cells, drain into larger vessels eventually draining into the subclavian veins.

B. Lymph flow is facilitated by one-way lymphatic valves and the pumping effect of skeletal muscle contraction.

1. The flow of lymph is assisted by contractions of smooth muscle associated with lymphatic vessel.

2. Functions of the lymphatic system include:
   a. Transporting fat from the gastrointestinal tract to blood.
   b. Filtering microorganisms and other foreign material from the lymph.

V. Regulation of Systemic Arterial Pressure

A. Cardiac output is determined by the end diastolic volume of the heart, which, in turn, is determined by the venous return.

1. The relationship between cardiac output (CO), mean arterial pressure (MAP), and total peripheral resistance (TPR) is described by the equation:

$$\mathbf{CO = MAP/TPR \text{ or } MAP = CO \times TPR}$$

a. Total peripheral resistance is the total resistance to the flow of blood from the aorta back to the right atrium.
b. Changes in peripheral resistance are entirely due to changes in the diameter of the arterioles and precapillary sphincters.

2. Cardiac output is sensitive to changes in end diastolic volume, which in turn depends on the total blood volume and the distribution of blood within the vascular system.

B. *Arterial baroreceptors are the sensing element of a negative feedback system that regulates heart rate so that mean arterial pressure is maintained relatively constant.*

1. **Baroreceptors** are located in the wall of the aortic arch and in the carotid sinus in the carotid arteries.

2. An increase in arterial pressure causes the baroreceptor to stretch, which initiates a transmission along the sensory neuron to the cardiovascular control center in the brain.

3. The pattern of action potential firing of stretch receptors in the baroreceptors reflects changes in both mean and pulsatile arterial pressures.

C. *The cardiovascular control center in the brain determines the level of automatic nerve activity to both the heart and the blood vessels.*

1. The cardiovascular control center for the baroreceptor reflex is located in the medulla.

2. It integrates inputs from the baroreceptors and other control centers in the brain.

D. *Cardiac output and total peripheral resistance are regulated by the cardiovascular control center so that mean arterial measure remains relatively constant.*

1. Changes in baroreceptor activity cause reciprocal changes in parasympathetic and sympathetic activity to the heart and blood vessels, which are innervated primarily by sympathetic nerves.

2. The overall response of the system to a change in blood pressure can be represented:
   a. Increased arterial pressure causes increased firing in the baroreceptor, which causes decreased sympathetic nerve activity and increased parasympathetic nerve activity to the heart and decreased cardiac output.
   b. Decreased sympathetic activity to arterioles leads to increased diameter in highly innervated vascular beds (skin, muscle), causing decreased peripheral resistance.
   c. **MAP = CO × TPR**
   d. Decreased TPR and decreased CO increased MAP

3. Increased arterial pressure and baroreceptor activity lead to a decrease in sympathetic activity to the veins and venodilation. This leads to a decrease in end-diastolic volume and cardiac output.

4. Regulation of blood volume is important for blood pressure regulation on a long-term basis.

5. Decreased cardiac output can be increased by:
   a. Increases in total circulatory blood volume.
   b. Shifting of blood volume from peripheral vascular blood to the heart by increasing venomotor tone.
   c. Increases in reabsorption of interstitial fluid by capillaries.

E. *Hemorrhage, hypotension, and exercise all cause primary changes in either cardiac output or total peripheral resistance that must be matched by compensatory changes elsewhere in the cardiovascular system to prevent large changes in mean arterial pressure.*

1. **Hemorrhage** causes a drop in arterial blood pressure that triggers the baroreceptor reflex and results in increase CO, TPR, and MAP.
   a. The vascular resistance to the heart and brain does not increase because of sparse sympathetic innervation.
   b. Selective innervation of different vascular beds provides a means for channeling blood to vital organs and away from nonvital organs to maintain homeostasis.
   c. With hemorrhage, stroke volume is reduced.

2. To compensate for hemorrhage and to increase blood volume, the following mechanisms operate:

a. Increase in venoconstriction.
b. Increase in absorption in capillary beds of interstitial fluid because of reduction of MAP and venous pressure.

3. **Hypotension,** or low blood pressure, results from inadequate filling of the heart and leads to decreased CO and inadequate flow of blood to the brain. Compensatory mechanisms include:
   a. A slight increase in sympathetic nerve activity.
   b. An increase in venomotor tone.
   c. An increase in CO.

4. During exercise, cardiac output and blood flow to skeletal muscles increase.
   a. Cardiovascular changes occurring during exercise are not mediated by the baroreceptor reflex.
   b. Local and humoral factors contribute to vasodilation of skeletal muscle vasculature.

5. Compensatory responses to exercise include:
   a. Increase in vascular resistance to skin, kidney, and gastrointestinal tract.
   b. Release of vasodilator substances in precapillary resistance vessels.
   c. Reflexive vasodilation of subcutaneous vessels.

VI. Cardiovascular Disease

A. Congestive heart failure refers to a variety of primary cardiac pathologies in which the heart is unable to maintain normal cardiac output.

1. **Heart failure** results from a reduction in cardiac contractility.
   a. Compensation for heart failure includes retention of water and sodium by the kidney to increase blood volume, end-diastolic volume, and CO.
   b. Compensatory mechanisms are limited and may lead to congestive heart failure.

2. Failure of the right ventricle leads to accumulation of blood in the right heart and veins, whereas failure of the left heart leads to accumulation of blood in the pulmonary circuit.

3. Treatment of heart failure includes antidiuretics, drugs to improve contractility, and angiotensin converting enzyme (ACE) inhibitors.

Key Terms

absorption
arterial pressure
arteriole
atherosclerosis
atrial natriuretic peptide
autoregulation
baroreceptor
capacitance vessel
capillary
cardiac output (CO)
cardiovascular control center
circulatory shock
colloid osmotic (oncotic) pressure
congestive heart failure
diastolic pressure
end diastolic volume
endothelial pores
hemorrhage
hydrostatic pressure
hypertension
hypotension
lymph
lymphatic system
MAP = CO × TPR
mean arterial pressure (MAP)
metarteriole
microcirculation
myogenic activity
precapillary sphincter
pulse pressure
Starling's hypothesis
systolic pressure
total peripheral resistance (TPR)
vasoactive center
vasoconstriction
vasodilation
vasomotion
venomotor tone
venous return
venule

Testing Your Knowledge

I. Matching. Select the correct term from the Key Terms.

_______ 1. difference between systolic and diastolic pressures

_______ 2. arterial pressure occurring during contraction of the ventricle

_______ 3. increase in contractile activity in response to stretch

_______ 4. ability of individual vascular beds to maintain constant blood flow

_______ 5. effect of increased carbon dioxide and decreased oxygen on arteriolar smooth muscle

_______ 6. effect of norepinephrine binding to alpha receptors or vascular smooth muscle

_______ 7. effect of epinephrine binding to beta receptors on smooth muscle

________ 8. pressure of blood owing to presence of plasma proteins

________ 9. describes balance of hydrostatic and oncotic pressures across capillary endothelium

________ 10. process favored when hydrostatic pressure is much greater than oncotic pressure

________ 11. process favored when precapillary sphincters contracted

________ 12. function as low pressure reservoir for blood

________ 13. structures sensitive to stretch

________ 14. functions to return fluid in interstitial space to the heart

________ 15. elevated arteriole pressure

II. Dichotomous. Circle the correct choice from the paired terms in parentheses.

1. The greatest drop of arterial pressure occurs at the level of the (arterioles/venules).

2. If the diameter of an arteriole is decreased by half, the resistance to blood flow (increases/decreases) by a factor of 16.

3. Increases in contractile activity of smooth muscle cells in arteriole walls in response to stretch results in a(n) (increase/decrease) in the diameter of the arteriole.

4. Norepinephrine binds to (alpha/beta/both) receptors whereas epinephrine binds to (alpha/beta/both) receptors on vascular smooth muscle cells.

5. The velocity of blood flow through capillaries is (inversely/directly) proportional to the cross-sectional area of the vessels through which it is flowing.

6. At the arterial end of a capillary, hydrostatic pressure is (greater/less) than oncotic pressure and leads to a net (filtration/absorption) pressure.

7. Your blood pressure suddenly increases from 90/60 to 160/90 mm Hg. The frequency action potentials by baroreceptors would (increase/decrease).

8. Your blood pressure drops suddenly because of hemorrhage. As a result, heart rate and contractility (increase/decrease) owing to (increased/decreased) sympathetic nerve activity to the heart.

9. During exercise, sympathetic activity to the heart is (increased/decreased), while parasympathetic activity is (increased/decreased).

10. During exercise potassium and adenosine cause (vasodilation/vasoconstriction) of precapillary resistance vessels.

III. True/False. In the space to the left of the number write T for true or F for false.

_______ 1. For a blood pressure of 120/70, the 120 represents diastolic pressure or pressure exerted during ventricular contraction.

_______ 2. In the water system analogy, the consumers are the organs.

_______ 3. Relaxation of smooth muscle cells in arteriole walls results in a decrease of arteriolar diameter.

_______ 4. You are running a marathon. The arterioles in your leg muscles would be dilated and those in your stomach would be constricted.

_______ 5. Vasodilation promotes filtration across the capillary endothelium, whereas vasoconstriction promotes absorption of fluid.

_______ 6. Blood pressure in venules and veins is greater than that in arterioles and arteries.

_______ 7. Lymphatic capillaries are identical to those in the circulatory system.

_______ 8. Plasma proteins in interstitial fluid can be returned to the blood only via the lymphatic system.

_______ 9. All vascular beds are innervated to the same degree of sympathetic innervation.

_______ 10. During mild to moderate exercise, the total peripheral resistance decreases as cardiac output increases.

IV. Sentence Completion. Write your response in the space provided.

1. Arterial pressure is lowest at the end of ventricular ________________.
2. If the diastolic pressure is 60 mm Hg and systolic pressure is 90 mm Hg, the pulse pressure is ________________.
3. Contraction of smooth muscle cells in the walls of arterioles ________________ the diameter of the vessel.
4. Bulk flow occurs in response to a _________ gradient between the inside and outside of the capillary, whereas diffusion occurs in response to a ______________ gradient.
5. In a capillary, plasma oncotic pressure is 28 mm Hg; tissue oncotic pressure is 3 mm Hg; and hydrostatic pressure is 35 mm Hg at the arteriole end, and 15 mm Hg at the venous end. The net filtration pressure is ______________ mm Hg at the arteriolar end, and the net absorption pressure is ______________ mm Hg at the venous end.
6. The cardiovascular control center is located in the ______________ of the brain.
7. End diastolic volume can be increased by increasing ______________ or by blood volume.
8. Failure of the ______________ ventricle leads to accumulation of fluid in the right heart and venous system.
9. The excess fluid accumulation in congestive heart failure may be reversed by the administration of ______________ , which enhance loss of sodium and water by the kidney.

V. Multiple Choice. Circle the letter of the best response.

1. Your pulse pressure is 90/60. Your mean arterial pressure is:
   a. 70 mm Hg
   b. 110 mm Hg
   c. 50 mm Hg
   d. 75 mm Hg
   e. 135 mm Hg

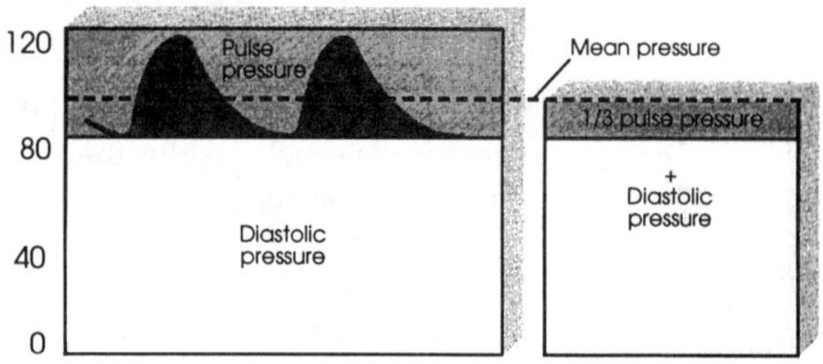

2. In the figure above, pressure is constant. If the radius of the outlet on the left increases by 4 compared to the radius of the outlet on the right, then:
   a. resistance on the left is half the resistance on the right
   b. flow on the left is greater than flow on the right
   c. flow on the left equals flow on the right
   d. flow on the left is less than flow on the right
3. In the water flow analogy, opening your spigot is analogous to:
   a. decreasing arteriole radius
   b. increasing local resistance to flow
   c. decreasing arteriole diameter
   d. decreasing local resistance to flow
   e. none of the above
4. You are being chased by angry bees. Which of the following is **not** occurring in your body?
   a. vasodilation of arterioles in your leg muscle by acetylcholine
   b. vasoconstriction of arterioles in your stomach by epinephrine
   c. vasodilation of arterioles in your heart by epinephrine
   d. binding of epinephrine to beta receptors on smooth muscle cells in your liver arterioles
   e. binding of epinephrine to alpha receptors of smooth muscle cells in your intestine arterioles
5. Compared to the venous end of the capillary, at the arterial end of a capillary:
   a. net absorption occurs
   b. plasma oncotic pressure is greater
   c. capillary hydrostatic pressure is greater

d. interstitial fluid colloid osmotic pressure is less
e. interstitial fluid hydrostatic pressure is greater

6. Which of the following events does **not** occur in response to a temporary increase in baroreceptor firing rate?
   a. decreased heart rate and contractility
   b. decreased total peripheral resistance
   c. increase in cardiac output and end diastolic volume
   d. vasodilation of heavily innervated vascular beds
   e. venodilation

7. All of the following occur in response to hemorrhage **except**:
   a. increased TPR
   b. venoconstriction
   c. increased fluid reabsorption from interstitial compartment
   d. erythropoiesis
   e. increased stroke volume

8. Fainting results primarily from:
   a. decreased CO
   b. increased SV
   c. increased vasoconstriction
   d. increased MAP
   e. decreased TPR

9. A decrease in cardiac contractility in heart failure leads to:
   a. decreased EDV
   b. decreased SV
   c. decreased CO
   d. decreased MAP
   e. all of the above

10. Accumulation of fluid in the pulmonary circuit occurs with:
    a. hemorrhage
    b. hypotension
    c. hypertension
    d. right heart failure
    e. left heart failure

VI. Complete the Table.

In the table, indicate the effect (vasoconstriction/vasodilation/no effect) of the factor on vascular smooth muscle.

Factor	Effect on Vascular Smooth Muscle
1. increased carbon dioxide	
2. decreased oxygen	
3. increased hydrogen ion concentration	
4. decrease in arterial pressure	
5. norepinephrine binding to alpha receptors	
6. epinephrine binding to beta receptors	
7. angiotensin II	
8. atrial natriuretic peptide	

VII. Complete the Table.

In the table, indicate the effect (increase/decrease/no effect) of the condition on arterial pressure and cardiac output.

Condition	Arterial Pressure	Cardiac Output
1. increased in venomotor tone		
2. increase in neural activity of baroreceptor		
3. decrease in neural activity of baroreceptor		
4. increase in end diastolic volume		
5. hemorrage		
6. hypertension		
7. decrease in sympathetic activity in the heart		
8. increase in sympathetic activity in the heart		

Chapter 20

Respiration

Chapter Learning Objectives

1. Describe the structure of the lungs, the airway tree, and the vascular tree.
2. Discuss airflow by comparing turbulent and laminar flows regarding noise, effect of diameter, and location in the respiratory system.
3. Explain vital capacity and ventilation and describe factors affecting each.
4. Specify the muscles used in breathing.
5. Indicate how pressure controls the movement of gases within the body.
6. Compare inspiration and expiration as to the role of muscles, changes in alveoli, pleural, and transpulmonic pressures and in the size of the thoracic cavity, alveoli, and lungs.
7. Describe how compliance is measured, what it represents, the effect of surface tension, and the medical significance of abnormally high and low values.
8. Describe respiratory diseases.

Expanded Chapter Outline

I. The Structure of the Lung

A. *Lung architecture is constrained by the principles of diffusion.*

1. Movement of gases in and out of cells occurs by simple diffusion.

2. All respiratory organs have three features in common:
   a. Very thin walls.
   b. Moist surfaces.
   c. High vascularity.

3. The surface area of lungs is increased by the formation of septa, whose walls are highly vascularized.

4. In the mammalian lung, septa evolved into microscopic air sacs called alveoli, each of which is surrounded by a capillary network.

B. *The **airway tree** divides repeatedly to enhance airflow.*

1. Each lung is composed of two tree-like structures, the vascular and airway trees, which are embedded in elastic tissue.

2. The flow of air through the airway tree is:

trachea → two bronchi (one to each lung) → bronchioles → alveoli, the exchange units

3. The conducting zone (trachea, bronchi, and bronchioles) has three important functions:
   a. To warm and humidify air.
   b. To distribute air to alveoli.
   c. To serve as part of the body defense system via a mucociliary transport system.

4. The respiratory zone (respiratory bronchioles, alveoli) functions in gas exchange.

II. Lung Volumes and Ventilation

A. *Respiratory muscles affect lung pressures.*

1. **Ventilation** is the process of moving air in and out of the lungs.

2. The key pressure-volume relationship in breathing is based on Boyle's Law: V is proportional to i/P, which states that the volume of a gas varies inversely with pressure at a constant temperature.

3. During **inspiration** the lungs play a passive role. Contraction of the **diaphragm** and external intercostal rib movements occur during normal inspiration.

4. Relaxation of the diaphragm occurs during

normal expiration, a passive process.

5. With exercise or forced expiration, the expiratory muscles, the abdominal muscles, and the internal **intercostal muscles** contract.

6. The total pressure of a mixture of gases is equal to the sum of the partial pressures of individual gases. The partial pressure of a gas is equal to the total pressure times the fractional concentration and is a measured in mm Hg.

7. Relative pressure is pressure relative to $\mathbf{P_B}$ **(barometric pressure)** and is measured in cm H_2O (1 cm = 0.74 mm Hg; 1 mm Hg = 1.36 cm H_2O). Pressures of breathing are relative pressures.

8. Three important relative pressures are associated with lung inflation and air flow:
   a. **Pleural pressure ($\mathbf{P_pl}$).**
   b. **Alveolar pressure ($\mathbf{P_A}$).**
   c. **Transpulmonic pressure ($\mathbf{P_{TP}}$)**, the gradient across the lung wall.

9. Transpulmonic pressure (P_{TP}), also called distending pressure, is equal to P_A minus P_{pl} and keeps the lungs from collapsing and allows them to stretch:

$$P_{TP} = P_A - P_{pl}$$

10. The sequence of events in inspiration is:

> contraction of inspiratory muscles → expansion of thoracic cavity → decrease in pleural pressure (becomes more negative) → increase in transpulmonic pressure (becomes more positive) → inflation of lungs → decrease in alveolar pressure (less than atmospheric) → flow of air into lungs

11. The sequence of events during expiration is:

> relaxation of respiratory muscles → decrease in volume of thoracic cavity → increase in pleural pressure (becomes less negative) → decrease in transpulmonic pressure → decrease in lung size → increase in alveolar pressure (greater than atmospheric) → flow of air out of lungs

B. *Lung volumes and airflow are measured by* ***spirometry.***

1. A spirogram measures the rate of airflow in and out of the lungs and the volume of air moved, the **tidal volume ($\mathbf{V_T}$).**

2. **Forced vital capacity (FVC),** the maximum amount of air forcefully and rapidly exhaled after a deep breath, is determined primarily by three factors:
   a. Strength of chest and abdominal muscles.
   b. Airway resistance.
   c. Lung volume.

3. **Forced expired volume ($\mathbf{FEV_1}$)** may be expressed as a percent of FVC that corrects for lung size differences.

4. **Residual volume** and **functional residual capacity** are measured indirectly by a dilution technique based on the relationship: $C_1 \times V_1 = C_2 \times V_2$, where C is concentration and V is volume.

C. *Some inspired air does not reach the alveoli.*

1. Minute ventilation equals tidal volume times breathing rate per minute.

2. **Dead space volume** is wasted air located in the conducting airways and is about 30% of tidal volume or minute ventilation.

3. Alveolar ventilation, the volume of fresh air available for gas exchange, is equal to the (tidal volume minus the dead space volume) times the breathing frequency.

4. Alveolar ventilation is affected by depth and rate of breathing.

III. Mechanical Properties of the Lung

A. ***Compliance*** *measures lung distensibility.*

1. The elasticity of the lungs affects how well they inflate and deflate with a change in transpulmonic pressure. The degree of lung expansion is proportional to PTP.

2. The elastic properties of a lung are determined by measuring changes in lung volume that occur with changes in pressure, plotted on a pressure-volume curve. The slope of the line is compliance, which provides a

measure of distensibility:

Compliance (C_L) = change in volume/change in pressure

3. Lungs with abnormally low compliance are difficult to inflate, whereas lungs with abnormally high compliance have poor elastic recoil.

4. Compliance is affected by surface tension occurring within the inner alveolar surface at the gas-liquid interface and produces a force that pulls inwardly.

5. Pulmonary **surfactant**, a complex of lipids and proteins, lowers surface tension, making the lung more compliant.

6. **Turbulent airflow** occurs at high flow rates and consists of disorganized patterns of airflow.

7. Turbulent flow increases resistance to breathing and is greatly affected by gas density.

8. **Laminar airflow** is characterized by streamlined flow that is parallel to the side of the airways.
   a. It is silent in contrast to turbulent flow.
   b. It is directly related to the radius raised to the fourth power **(Flow ∞ radius^4).**

9. The amount of work involved in ventilating the lungs is affected by the degree of:
   a. Elasticity of the lung.
   b. Surface tension in the alveoli.
   c. Airway resistance.

Key Terms

airway tree
alveolar pressure (P_A)
alveolar ventilation
alveolus
bronchii
compliance
conducting unit
dead space volume
diaphragm
expiration
forced expiratory volume (FEV_1
pneumothorax
forced vital capacity (FVC)
functional residual capacity
hyperventilation
hypoventilation
inspiration
intercostal muscle
laminar flow
minute ventilation
pleural pressure (P_{pl})
pleural space
pressure-volume curve
pulmonary surfactant
relative pressure
residual volume (RV)
respiration
spirometry
surface tension
thoracic cavity
tidal volume
transpulmonic pressure (P_{TP})
turbulent flow
vascular tree
ventilation

Testing Your Knowledge

I. Matching. Select the correct term from the Key Terms.

________ 1. gas exchange area of lung

________ 2. process of moving air into and out of lungs

________ 3. type of airflow in trachea and bronchi

________ 4. amount of air entering or leaving lungs during a single breath

________ 5. maximum amount of air forcefully and rapidly exhaled after maximal inspiration

________ 6. air remaining in lungs after forced expiration

________ 7. volume of air remaining in lungs after normal expiration

________ 8. amount of air in conducting units

________ 9. dome-shaped muscle that contracts during inspiration

________ 10. pressure gradient across lung wall

________ 11. condition caused by lung collapse

________ 12. pressure became subatmospheric during inspiration

________ 13. size or capacity increases during this process

________ 14. determined by slope of pressure-volume curve

________ 15. reduces surface tension.

II. Dichotomous. Circle the correct choice from the paired terms in parentheses.

1. There is a(n) (direct/inverse) relationship between alveolar diameter and metabolic rate.

2. Pleural pressure becomes (more/less) negative during inspiration but (more/less) negative during expiration.

3. Transpulmonic pressure is more affected by changes in (pleural/alveolar) pressures.

4. Residual volume is measured at the end of (forced/normal) expiration.

5. Lung compliance and distensibility are (directly/inversely) related.

6. Given two lungs, A and B, with the same ΔV on a lung pressure-volume curve, if ΔP for lung A is less than that for lung B, the compliance of lung A will be (less/greater) than that for lung B.

7. In a healthy individual, the (residual/dead space) volume is about 20% of the total lung capacity.

8. A person with (asthma/emphysema) would have a FEV/FVC ratio greater than or equal to 0.8.

9. A person with (bronchitis/emphysema) would have a greater lung compliance value than a healthy person.

10. Surface tension has a greater effect on (smaller/larger) alveoli.

III. True / False. To the left of the number write T for true or F for false.

_______ 1. A relative pressure of minus 5 cm H_2O is five times lower than the atmospheric pressure.

_______ 2. Transpulmonic pressure increases in a collapsed lung.

_______ 3. During exercise, pleural pressure becomes more positive.

_______ 4. Functional residual capacity is greater than the residual volume.

_______ 5. If the total lung capacity is 6 L and the residual volume is 1.2 L, the vital capacity is 4.8 L.

_______ 6. To attain maximum alveolar ventilation, you should breathe more rapidly.

_______ 7. The residual volume of air is that found in the conducting zone.

_______ 8. Gas exchange occurs in the alveoli and bronchi.

_______ 9. There is a direct relationship between the size of a mammal, its metabolic rate, and its alveolar diameter.

_______ 10. Particles less than 5 μm generally get inhaled into the lung.

IV. Sentence Completion. Write your response in the space provided.

1. Alveolar pressure at midinspiration is _____________ than atmospheric pressure.

2. Given a $P_A = -1$ and $P_{TP} = 6$, the P_{pl} is _____________.

3. Given a $P_{pl} = -6$ and $P_A = +1$, the P_{TP} is _____________.

4. A person with a FVC of 3.8 L and FEV_1 of 3.04 L would have a(n) _____________ lung disorder.

5. A person with a FEV_1 of 2.8 L and FVC of 3.8 L would have a(n) _____________ lung disorder.

6. Given a tidal volume of 500 mL and minute ventilation of 8 L/minute, the breathing rate is _____________.

7. If the tidal volume increases, the ratio of dead space volume to tidal volume _____________.

8. Given a V_T of 600 mL and breathing ratio of 15 breaths/minute, the minute volume is _____________.

9. Given an IRV of 3.1 L, VT of 600 mL and ERV of 1.2 L, the vital capacity is _________.

10. Given vital capacity of 4.9 L and residual volume of 1100 mL, the TLC is __________.

V. Multiple Choice. Circle the letter of the best response.

1. All the following occur during inspiration except:
   a. contraction of diaphragm
   b. contraction of external intercostals
   c. increase in thoracic cavity volume
   d. decrease in pressure in the pleural space
   e. positive relative pressure

2. Transpulmonic pressure is greatest at the:
   a. end of expiration
   b. middle of inspiration
   c. end of inspiration
   d. middle of expiration
   e. beginning of inspiration

3. When air flows out of the alveoli through the airways:
   a. alveolar pressure is less than atmospheric pressure
   b. alveolar pressure is equal to 0
   c. pleural pressure is greater than atmospheric pressure
   d. transpulmonic pressure decreases
   e. pleural pressure is less than transpulmonic pressure

4. In a person with asthma:
   a. FCV is greater than FEV_1
   b. FCV equals FEV_1
   c. FCV/FEV_1 is less than 0.8
   d. FEV_1/FVC is much less than 0.8
   e. FEV_1/ FVC is greater than 0.8

5. In a healthy person alveolar ventilation may be increased by:
   a. increasing breathing rate and dead space ventilation
   b. increasing tidal volume and breathing rate
   c. decreasing breathing rate and increasing tidal volume
   d. decreasing breathing rate alone
   e. increasing breathing rate alone

6. Which volume or measurement would probably be the same in person A with emphysema and in person B with asthma?
   a. FEV_1
   b. V_T
   c. compliance
   d. dead space volume
   e. FVC

7. For which parameter would you expect to see a difference between a normal person and someone with asthma?
   a. compliance
   b. tidal volume
   c. alveolar pressure
   d. FVC
   e. pleural pressure

8. The flow of air into the alveoli is:
   a. trachea bronchioles → bronchi → alveoli
   b. trachea → bronchi → bronchioles → alveoli
   c. bronchi → trachea → bronchioles → alveoli
   d. bronchioles → trachea → bronchi → alveoli

9. In a normal person, the FEV1 is:
   a. 6.0 L
   b. 3.8 L
   c. 4.8 L
   d. 2.4 L
   e. 3.1 L

10. All of the following can be determined from a spirogram except:
    a. TLC
    b. RV
    c. FEV1
    d. IRV
    e. VC

VI. Complete the Table.

Indicate what happens to the pressure or structure during inspiration and expiration.

Pressure or Structure	Inspiration	Expiration
1. diaphragm		
2. lung capacity		
3. thoracic capacity		
4. pleural pressure		
5. alveolar pressure		
6. transpulmonic pressure		

Chapter 21

The Pulmonary Circulation and Gas Uptake

Chapter Learning Objectives

1. Give the details of the structure of pulmonary circulation.
2. Explain how blood pressures in the heart and pulmonary circulation may be analyzed and the effect of exercise on pulmonary blood pressure determined.
3. Explain ventilation-perfusion balance.
4. Account for whole lung hypoxia.
5. Trace the diffusion of gases in the alveoli and the alveolar-capillary membrane.
6. Cover the uptake and transport of oxygen by the blood; describe the role of hemoglobin in the transport of oxygen and draw an oxyhemoglobin dissociation curve.
7. Describe the mechanism involved in carbon dioxide transport.
8. Characterize the two types of pulmonary edema.
9. Describe fetal circulation.
10. Indicate the changes in pulmonary circulation at birth.
11. Explain the controls over breathing.
12. Account for some of the abnormal breathing patterns.

Expanded Chapter Outline

I. Functional Design of the Pulmonary Circulation

A. Pulmonary vascular tree accepts all of the cardiac output.

1. The branching of the vascular tree parallels that of the airway tree.
2. Capillaries are embedded in alveolar walls. Capillaries from more than one alveolus join to form capillary beds to provide different pathways across the gas exchange surface.
3. Lymph vessels drain excess fluid from the interstitial spaces between alveoli and capillaries and return it to systemic veins.

B. Pulmonary circulation has metabolic functions.

1. Functions of endothelial cells in the capillary wall include:
   a. Conversion of **angiotensin I** to active **angiotensin II**, a vasoconstrictor.
   b. Release and inactivation of other vasoactive substances such as histamine, serotonin, and prostaglandins.
2. Angiotensin II causes constriction of arterioles to raise blood pressure.

II. Vascular Pressure and Blood Flow

A. Pulmonary circulation is a low pressure system.

1. Pulmonary circulation is a low pressure, low resistance system with highly compliant vessels. Pulmonary arterial diastolic pressure is about 10 mm Hg.
2. Mean pulmonary arterial pressure for a normal resting adult is 15 mm Hg.
3. Most of the blood flows through the lower lung.

B. Hypoxic vasoconstriction balances blood flow with ventilation.

1. Uneven distribution of inspired air to the alveoli results in regional **hypoxia**.
2. Through a mechanism called **hypoxic vasoconstriction**, the lung balances the flow in perfusion of blood with the availability of regional alveolar oxygen and restricts blood flow to hypoxic areas.
3. Whole lung hypoxia leads to a rise in pulmonary arterial pressure, which forces blood to the upper regions of the lungs,

where oxygen is relatively abundant.

C. Exercise recruits capillaries and decreases transit time.

1. As cardiac output and pulmonary arterial pressure increase, more blood flows to the upper lung, where reserve capillaries are recruited.

2. As cardiac output increases, transit time through pulmonary capillaries decreases, which allows more blood to be oxygenated every second during exercise.

III. Gas Uptake and Transport

A. Partial pressures move respiratory gases across the alveolar-capillary membrane.

1. The partial pressure of oxygen (P_{02}) in air is 160 mm Hg as derived from Dalton's law: $\mathbf{P_{02} = P \times F_2}$ (PO_2 = partial pressure of O_2; PB = atmospheric pressure; FO_2 = fractional concentration of O_2)

2. P_{02} in moist tracheal air at 37° C is 149.2 mm Hg and P_{02} in alveoli is approximately 100 mm Hg.

3. The oxygen gradient favors movement of oxygen from alveolar gas into capillary blood, whereas the reverse is true for carbon dioxide.

4. Oxygen dissolves in the capillary membrane then passes into the plasma and through the erythrocyte membrane to hemoglobin.

B. Oxygen is transported by hemoglobin.

1. At rest, 250 mL/minute of oxygen is taken up through pulmonary capillaries and transported to tissue capillaries. During heavy exercise, 3750 mL/minute of oxygen are distributed through pulmonary capillaries.

2. Each hemoglobin molecule combines reversibly with four oxygen molecules to form oxyhemoglobin.
   a. In tissues hemoglobin releases oxygen.
   b. The capacity of hemoglobin for oxygen is much higher than that of plasma.

3. The maximum amount of O_2 that can be carried by hemoglobin is called oxygen carrying capacity and is expressed as Vol%. The percent saturation of hemoglobin is the ratio of O_2 content x 100/O_2 carrying capacity = percent O_2 saturation, where O_2 content is the amount actually bound.

4. The **oxyhemoglobin dissociation curve** shows the way in which oxygen binds to hemoglobin in the lung and unloads oxygen in the tissues. Under normal conditions the partial pressure of oxygen in arterial and venous bloods is 100 mm Hg and 40 mm Hg, respectively.

5. Under normal conditions, arterial blood O_2 saturation is 97% and venous blood is 75%.

6. Heavy exercise enlarges the oxygen gradient, which favors the movement of oxygen from capillary blood to the tissue.

7. Hemoglobin unloads oxygen more readily under conditions of acid pH and increased temperature. The oxyhemoglobin dissociation curve shifts to the right. A leftward shift of the oxyhemoglobin dissociation curve enhances oxygen loading and occurs when the pH is more basic or the temperature drops.

C. Carbon dioxide is transported as a bicarbonate ion.

1. The largest fraction (60% to 70%) of CO_2 is transported as bicarbonate ions, which are formed inside erythrocytes by the following reaction series:

$$CO_2 + H_2O + \text{carbonic anhydrase} \rightarrow H_2CO_3 \rightarrow HCO_3^- + H^+ + Hb^- \rightarrow HHb$$

2. As bicarbonate ions (HCO_3^-) move into the plasma and the H^+ combines with hemoglobin, chloride ions move into the erythrocyte, the chloride shift.

3. CO_2 can combine loosely with hemoglobin at a different binding site on hemoglobin than oxygen to form carbaminohemoglobin.
   a. Between 15% ≈ 30% of CO_2 is transported in this way.
   b. Hemoglobin has a higher affinity for CO_2 when it is oxygen-desaturated.

4. Between 7% ≈ 10% of CO_2 is transported in a dissolved state in the plasma and erythrocytes.

5. Pulmonary **edema** results from the accumulation of fluid in the interstitial spaces and increases the thickness of fluid through which oxygen must diffuse.

IV. Fetal and Neonatal Circulatory Physiology

A. The placenta is the gas-exchange organ before birth.

1. In the fetus blood is shunted away from the lungs by two passages: the **foramen ovale** between the right and left atria and the **ductus arteriosus** between the pulmonary artery and aorta.

2. At birth the foramen ovale and ductus arteriosus close so blood passes from the right atrium to the right ventricle and through the pulmonary artery to the lungs.

B. Congenital heart disorders interfere with gas exchange.

1. Failure of the closure of the ductus arteriosus can lead to pulmonary hypertension and right ventricular failure.

2. Atrial and ventricular septal defects result from incomplete formation of the septa between the atria and between the ventricle.

V. The Control of Breathing

A. Breathing has a basic rhythm.

1. **Inspiratory neurons** are responsible for the basic rhythm and activate the inspiratory muscles. Neuronal inputs from the pons modulate the inspiratory center and act as a "cut-off" signal to inhibit inspiration in automatic breathing.

2. **Pulmonary stretch receptors** in the smooth muscles of the airways send messages to the inspiratory neurons to inhibit inspiration and prevent overinflation.

3. **Expiratory neurons** in the medulla activate expiratory muscles during forceful breathing.

4. The respiratory center can be disrupted by cerebral concussions and pressures in addition to drugs such as barbiturates.

B. Changes in blood gases stimulate ventilation.

1. The rate and depth of breathing vary depending on metabolic demands.

2. Oxygen, carbon dioxide, and hydrogen ions affect both the rate and the depth of breathing.

3. Central chemoreceptors in the medulla monitor changes in cerebral arterial carbon dioxide levels.

4. Peripheral chemoreceptors in the aortic arch and at the bifurcation of the common carotid artery monitor levels of 0_2, CO_2 and H^+ in arterial blood.
   a. Sensory neurons from these receptors extend to the respiratory center in the medulla.
   b. Arterial CO_2 has the greatest effect on ventilation.
   c. An increase in CO_2 causes increased alveolar ventilation, whereas a decrease in CO_2 causes hypoventilation.
   d. The control of ventilation by CO_2 is mediated by central and peripheral chemoreceptors.
   e. The control of ventilation by H^+ occurs exclusively through peripheral chemoreceptors.
   f. A change in arterial H^+ is caused by altered blood CO_2 and other factors.

8. Oxygen has the smallest effect on ventilation.

C. Breathing patterns can be abnormal:

1. Sleep **apnea** involves obstruction of the upper airways during sleep.

2. Shallow-water blackout.

3. The diving reflex.

4. The pickwickian syndrome.

Key Terms

alveolar-capillary membrane
angiotensin I and II
aortic bodies
atrial septal defect
bicarbonate ion
capillaries
carbaminohemoglobin
carbonic anhydrase
cardiac catheterization
carotid bodies
ductus arteriosus
embolus
expiratory neuron
foramen ovale
hypoxemia
hypoxiasleep apnea
inspiratory neuron
oxyhemoglobin
oxyhemoglobin dissociation curve
pulmonary artery
pulmonary edema
pulmonary hypertension
pulmonary stretch receptor
pulmonary vein
ventricular septal defect

Testing Your Knowledge

I. Matching. Select the correct term from the Key Terms.

________ 1. carries blood from the right ventricle to the lungs

________ 2. accumulation of fluid in the lung

________ 3. vasoconstrictor converted to active form by endothelial cells

________ 4. describes low oxygen concentration in the lung

________ 5. condition of low blood oxygen

________ 6. condition of elevated arterial pressure in the lungs

________ 7. form in which most $C0_2$ transported in blood

________ 8. molecule formed from binding of $C0_2$ to hemoglobin

________ 9. blood clot

________ 10. opening between right and left atria

II. Dichotomous. Circle the correct choice from the paired terms in parentheses.

1. Blood flows from the heart through the pulmonary (artery/vein) to the lungs.
2. Hypoxia occurs more frequently in alveoli located in the (upper/lower) part of the lung.
3. At an altitude of 6000 feet the PO_2 in alveolar gas is (greater/less) than the PO_2 in alveolar gas at sea level.
4. As the number of erythrocytes increases, the O_2 carrying capacity (increases/decreases).
5. Under (acidic/basic) conditions, the oxyhemoglobin dissociation curve shifts to the right to facilitate oxygen (loading/unloading).
6. In a fetus the oxyhemoglobin dissociation curve would shift to the (left/right).
7. Given an arterial PO_2 of 100 mm Hg, the venous PO_2 would be (higher/lower).
8. Given an arterial PCO_2 of 40 mm Hg, the venous PCO_2 would be (higher/lower).
9. The vol% for a P_{CO2} of 40 mm Hg is (greater/less) that the vol% for a PO_2 of 40 mm Hg.
10. Under normal breathing, expiration is a(n) (active/passive) process and (does/does not) require neuronal input from the expiratory neurons.

III. True/False. In the space to the left of the number write T for true or F for false.

______ 1. Blood flow is restricted to hypoxic alveoli.

______ 2. The P_{CO2} of mixed venous blood is less than the P_{CO2} in alveolar gas.

______ 3. The oxygen carrying capacity in a person with anemia is the same as that in a healthy person.

______ 4. The oxyhemoglobin dissociation curve is shifted to the right under acidic conditions.

______ 5. About 60% ≈ 70% of CO_2 is carried as carbaminohemoglobin.

______ 6. The respiratory center is located in the pons.

_______ 7. The carotid bodies monitor levels of O_2, CO_2, and H^+.

_______ 8. An increase in arterial P_{CO2} can cause an increase in H^+ concentration.

_______ 9. H^+ can diffuse across the blood-brain barrier, but CO_2 cannot.

_______10. Ventilation may be inhibited by low H^+ concentration, increased P_{CO2} and decreased PO_2.

IV. Sentence Completion. Write your response in the space provided.

1. Under a condition of whole lung hypoxia, pulmonary arterial pressure increases as a result of ____________, but during exercise pulmonary arterial pressure increases as a result of ____________.
2. Oxygen tension in alveoli is approximately ____________ at sea level.
3. The oxygen carrying capacity of a person with a hemoglobin of 12 g/100 mL is ____________.
4. Given an O_2 carrying capacity of 20 vol% for hemoglobin and % saturation of 50%, the O_2 content is ____________.
5. A leftward shift of the oxyhemoglobin curve facilitates oxygen ____________.
6. The O_2 content and percent saturation are higher at a given O_2 tension when the oxyhemoglobin dissociation curve shifts ____________.
7. If the ductus arteriosus does not close after birth, blood may be shunted between the and the ____________.
8. Inspiration is inhibited by neuronal input from the ____________ in the brain and ____________ in the airways.
9. Central chemoreceptors located in the ____________ monitor changes in ____________.
10. Short-term regulation of P_{CO2} in the blood is through ____________.

V. Multiple Choice. Circle the letter of the best response.

1. Pulmonary arterial pressure is highest during/at:
   a. rest
   b. moderate exercise
   c. exercise at 8000 feet above sea level
   d. maximal exercise at sea level
   e. an elevation of 8000 feet above sea level

2. During exercise:
   a. hypoxic vasoconstriction occurs in the lungs
   b. pulmonary arterial pressure increases
   c. blood flow through capillaries in the lungs is decreased
   d. capillary diameter increases
   e. the perfusion rate of alveoli in the upper lungs increases

3. Oxygen movement from alveoli to hemoglobin occurs by:
   a. diffusion
   b. active transport
   c. facilitated diffusion
   d. exocytosis
   e. pinocytosis

4. Under which condition would oxygen unloading be facilitated?
   a. increased pH
   b. decreased temperature
   c. increased P_{CO2}
   d. increased P_{CO2}
   e. all of the above

5. In the fetal heart, blood is shunted between:
   a. right and left atria
   b. right atrium and left ventricle
   c. left atrium and right ventricle
   d. right and left ventricles
   e. right atrium and pulmonary artery

6. If the foramen ovale or ductus arteriosus fails to close after birth, the result may be
   a. pulmonary hypertension
   b. hypoxemia
   c. a decreased arterial PO_2
   d. right ventricular failure
   e. all of the above

7. Which of the conditions is the most powerful stimulant to alveolar ventilation?
   a. increased PO_2
   b. decreased PO_2

c. increased H^+ concentration
d. decreased arterial pH
e. increased P_{CO2}

8. Under which condition would alveolar ventilation be decreased?
a. decreased P_{CO2}
b. decreased PO_2
c. increased P_{CO2}
d. increased H^+ concentration
e. increased arterial blood pH

9. Under which condition(s) is alveolar ventilation stimulated via peripheral chemoreceptors?
a. increase in arterial H^+ concentration
b. decrease in PO_2
c. hyperventilation
d. breathing at a high altitude
e. all of the above

10. Which abnormal breathing pattern is characterized by low P_{CO2}, low PO_2, hyperventi lation, and hypoxemic brain?
a. sleep apnea
b. Cheyne-Stokes breathing
c. shallow-water blackout
d. diving reflex
e. pickwickian syndrome

VI. Complete the Table.

For each condition, indicate the direction (right or left) an oxyhemoglobin dissociation curve would shift, and whether loading or unloading of oxygen is facilitated.

Condition	Shift (right, left)	Loading/Unloading
1. presence of (H^+)		
2. low P_{O2}		
3. elevated temperature		
4. more basic pH		
5. high (CO_2)		

VII. Complete the Table.

For each condition, indicate the effect (increase/decrease/no effect) on ventilation.

Condition	Ventilation
1. increase in arterial CO_2	
2. decrease in arterial O_2	
3. increase in (H^+)	
4. decrease in arterial CO_2	
5. decreased blood pH	
6. vomiting	

Chapter 22

The Gastrointestinal System

Chapter Learning Objectives

1. Describe the anatomy, processes (motility, secretion, digestion, and absorption), and regulation of the gastrointestinal system.
2. Describe neural control of the gastrointestinal processes of motility and secretion by short and long reflexes.
3. Give the function of chewing and peristaltic motion. Describe the composition and role of saliva.
4. Follow the passage of food through the organs of the digestive tract.
5. Indicate the source and function of the hormones (gastrin, cholecystokinin, secretin, and gastric inhibitory peptide) that participate in control over the gastrointestinal tract.
6. Discuss the roles of the various secretions of the pancreas and how pancreatic juice secretion is regulated.
7. Describe the function of the gallbladder, the formation of gallstones, and the composition of bile, and discuss how its secretion is regulated.
8. Indicate what happens to the food in the stomach and small intestine. Describe the mechanisms by which the carbohydrates, proteins, and triacylglycerides are digested. and absorbed into epithelial cells and then to the blood or lymphatic system.
9. Describe the mechanisms for the absorption of water and inorganic ions from the small intestine.
10. Describe what happens in the large intestine. Describe the types of molecules and mechanisms by which these molecules are absorbed in the large intestine.
11. Describe the types of motility in the large intestine and the defecation reflex.

Expanded Chapter Outline

I. Overview of the Gastrointestinal System

A. *Anatomical features of the gastrointestinal system are the digestive tract and its accessory structures.*

1. The digestive tract consists of the mouth, pharynx, esophagus, stomach, and small and large intestines. Accessory structures include salivary glands, pancreas, liver, and gallbladder.

2. The wall of the digestive tract is composed of four coats:
   a. **Mucosa**, which consists of the epithelial lining, lamina propria, and muscularis mucosa.
   b. **Submucosa**, which consists of a layer of connective tissue with lymph and blood vessels, exocrine glands, and a plexus of nerve cells.
   c. **Muscularis externa**, which consists of circular and longitudinal muscle layers and a network of nerve cells.
   d. **Serosa**, which consists of a layer of connective tissue.

3. The salivary glands and pancreas contain clusters of exocrine secretory cells called the **acini**, which lead into a system of ducts that converge and empty into the mouth or duodenum.

4. The circulation of the stomach, small and large intestines, pancreas, and liver is called the **splanchnic circulation**. The **portal circulation** brings blood from the stomach, intestines, and pancreas to the liver, where digestive products are metabolically processed.

B. *The processes of the gastrointestinal system are carried out by various cell types.*

1. The motor functions of most of the digestive tract are performed by smooth muscle layers.

2. The two basic types of motor activity involved in the digestive and absorptive functions are:
   a. Mixing movements.
   b. Propulsive movements whose rate

depends on the function served by the organ.

3. Important molecular components of the gastrointestinal system include:
   a. Inorganic ions such as hydrogen and bicarbonate that are secreted by the stomach and pancreas, respectively.
   b. Organic components such as digestive enzymes of the salivary glands, stom ach, and pancreas; bile salts, pigments, and cholesterol, which are secreted by the liver; and mucin, which is produced by all regions of the digestive tract.
   c. Water and solutes that are transferred across secretory membranes by a variety of active and passive transport processes.

4. Through enzyme digestion, large organic molecules are broken down into smaller molecules that can be absorbed.

5. Various transport systems are used to move digestive products from the lumen to the blood and lymph.

C. *Gastrointestinal function is regulated by neural and hormonal mechanisms.*

1. Motility and secretion are regulated by neural and hormonal mechanisms, while digestion and absorption are not.

2. Many of the stimuli that initiate the regulatory events are present in the lumen of the digestive system.

3. Chemoreceptors and mechanoreceptors respond to the stimuli.

4. Two types of neural pathways are involved in the regulation of gastrointestinal motor and secretory activity:
   a **Short reflexes** mediated by nerves lying within the wall of the digestive tract.
   b. **Long reflexes** mediated by nerves outside of the tract.

5. Short reflexes can be excitatory or inhibitory, depending on the mediator released at the effector cell by motor and secretory neurons.

6. Long reflexes coordinate activities of different regions of the gastrointestinal system and between the system and other body parts: Long reflexes involve the central and autonomic nervous systems.

7. Four hormones important in the control of motility and secretion are gastrin, cholecystokinin (CKK), secretin, and gastric inhibitory peptide (GIP).

8. These hormones produced by endocrine cells in the epithelial lining of the tract are released in response to intraluminal stimuli acting directly on the endocrine cells.

II. Chewing, Salivary Secretion, and Swallowing

A. *Chewing reduces the size of food particles and mixes them with saliva.*

1. Chewing is accomplished by combined actions of the skeletal muscles of the jaws, lips, cheeks, and tongue.

2. Chewing is partly reflexive in nature and aids digestion.

B. *The long reflex secretion of saliva lubricates the mouth and initiates the digestion of starch.*

1. Saliva is secreted by the **parotid, submaxillary**, and **sublingual glands** in response to chewing and is composed of water, mucus, and amylase.

2. Secretion of saliva is regulated by long reflexes involving sympathetic and parasympathetic nerves.

3. Salivary secretion serves a number of functions, including:
   a. Softening and lubricating food.
   b. Neutralizing acids produced by oral bacteria.

4. Salivary amylase degrades complex carbohydrates.

C. *Swallowing moves food rapidly from the mouth into the stomach.*

1. Long reflexes mediated through a swallowing center in the brain coordinate actions of smooth and skeletal muscles involved in swallowing the **bolus** of food.

2. Food is prevented from entering the **glottis** by the **epiglottis**.

3. A bolus of food moves from the pharynx through the relaxed upper sphincter into the esophagus in response to the downward pressure gradient existing through these regions.

4. A wave of contraction called **peristalsis** along the esophagus moves the bolus to the stomach through the relaxed lower sphincter.

5. The **all-or-none swallowing reflex** is activated when a bolus stimulates mechanoreceptors in the pharynx.

III. The Processes of Motility and Secretion in the Stomach

A. *The muscular activities of the stomach associated with gastric filling and emptying are regulated by neural and hormonal mechanisms.*

1. The motor activity of the stomach serves the major functions of:
   a. Mixing food with gastric secretions.
   b. Conversion of large solid food particles to liquid suspension of fine particles.
   c. Emptying of gastric contents into the duodenum.
   d. Storage.

2. The storage function is served by the smooth muscle of the stomach.
   a. The plasticity of smooth muscle allows the stomach to accommodate large volumes with minimal changes in intragastric pressure.
   b. Peristaltic contractions mix food with gastric juice, reduce particle size, and expel gastric contents into the duodenum.

3. The stomach has a basic electrical rhythm that consists of cycles of spontaneous depolarization and repolarization, originating in muscle pacemaker cells at the top of the stomach.
   a. When excitatory stimuli are sufficient to produce action potentials, contraction becomes more vigorous.
   b. Neural and hormonal mechanisms are involved in regulating the rate of gastric emptying.
   c. Stretching of stomach mechanoreceptors and the presence of gastrin increase gastric motor activity.

4. Inhibitory mechanisms prevent the small intestine from being overwhelmed by material from the stomach.
   a. The mechanisms originate in the duodenum.
   b. **Secretin** and **gastric inhibitory hormone,** released by the duodenum mucosa when contacted by acid and fat, respectively, inhibit gastric motor activity.

5. The **enterogastric inhibitory reflex,** which is initiated by stimulation of receptors in the duodenum, also inhibits gastric motility.

B. *Gastric secretion of hydrochloric acid is by an active transport system and is regulated by neural and hormonal mechanisms.*

1. The exocrine secretory glands contain three types of secretory cells:
   a. **Chief cells,** which secrete pepsinogen.
   b. **Parietal cells,** which secrete HCl and intrinsic factor, a substance necessary for vitamin B_{12} absorption.
   c. **Mucous cells,** which produce alkaline mucus that protects the mucosa.

2. Hydrogen ions are actively transported into the gastric lumen. Chloride ions are actively transported from blood to parietal cells and then to the gastric lumen.

3. Excitation and inhibition of secretion occur in three phases: the **cephalic, gastric, and intestinal phases**. Both short and long reflexes via the vagus nerve are involved.

4. **Pepsinogen,** the precursor of pepsin, is synthesized and stored in the chief cells until it is released into the gastric lumen. There it is converted to **pepsin** by HCl.

5. Secretion of pepsinogen involves short and long reflexes that occur during the cephalic and gastric phases of digestion.

6. Both peptides and acetylcholine play important roles in the activation of pepsin and in the maintenance of the highly acidic environment required for optional activity of pepsin.

C. *Vomiting rapidly expels gastric contents through the mouth.*

1. **Vomiting** is a reflex act coordinated by a center located in the medulla.

2. The force for expulsion comes from an increase in intragastric and abdominal pressure due to contraction of the diaphragm and abdominal muscles.

3. Vomiting may be triggered by emetics, disturbances in the vestibular apparatus, and stimulation of sensory nerve endings in the throat.

IV. The Process of Secretion in the Exocrine Pancreas

A. The secretion of pancreatic bicarbonate is by an active transport system.

1. The epithelial cells lining the intercalated ducts secrete an alkaline juice that contains a high concentration of sodium bicarbonate.

2. The bicarbonate protects the duodenum from damage by gastric acid and provides an optimum pH for pancreatic digestive enzymes.

B. The pancreas secretes a wide variety of digestive enzymes.

1. The pancreas secretes three protein-digesting enzymes: **trypsin**, **chymotrypsin**, and **carboxypeptidase**.

2. **Pancreatic lipase** digests triacylglycerol fats.

C. Regulation of pancreatic bicarbonate and enzyme secretion is by hormonal mechanisms.

1. The hormones secretin and cholecystokinin, released from the duodenal mucosa during digestion, play the central role in the control of secretion by the pancreas.

2. **Secretin** and **cholecystokinin** are released in response to acid, protein, and fat digestion products, respectively.

V. The Processes of Secretion in the Biliary System

A. The gallbladder stores and concentrates bile secreted by the liver.

1. **Bile**, secreted by hepatocytes and stored in the gallbladder, plays a role in the digestion and absorption of fat.

2. Bile is composed of bile salts, pigments, cholesterol, lecithin, and inorganic ions such as sodium, chloride, and bicarbonate.

3. In the gallbladder, the organic constituents of bile become concentrated by the removal of sodium and fluid.

B. Bile salts are recycled between the liver and small intestine.

1. **Bile salts**, synthesized from cholesterol by hepatocytes, play an essential role in the digestion and absorption of fat.

2. Bile salts are absorbed from the small intestine, returned to the liver, and resecreted in bile. This recycling is called the **enterohepatic circulation of bile salts**.

C. Regulation of the secretion of bile is by the portal blood concentration of bile salts and by secretion.

1. Bile secretion is regulated by the concentration of bile salts in the portal blood.

2. Secretin stimulates epithelial cells lining the bile ducts to release sodium bicarbonate to neutralize gastric acid in the small intestine.

VI. The Processes of Motility, Secretion, and Absorption in the Small Intestine

A. The muscular movements of the small intestine are segmenting contractions (mixing movements) and peristalsis.

1. **Segmenting contraction**, the most frequently occurring type of muscular movement, moves the contents of different regions of the small intestine back and forth over small distances.
   a. This action mixes the contents with digestive secretions.
   b. The mixing brings products of digestive activity into contact with the absorptive epithelium.

2. **Pacemaker** cells determine the frequency of these movements.

3. Peristaltic movements provide slow transit required for completion of the absorptive processes.

4. Chemical and mechanical stimulation of

receptors in the intestinal wall initiate long and short reflexes, which are important in regulation of small intestinal mobility.

5. **Gastrin** increases motor activity of the ileum and relaxes the **ileocecal sphincter**.

B. Water and solute move from blood to small intestinal lumen in response to osmotic and solute concentration gradients and by secretion.

1. Fluid moves from blood to the intestinal lumen in response to the hyperosmotic contents of the lumen.

2. Mucosal exocrine cells secrete water, inorganic ions, and mucus into intestinal lumen.

C. Absorption of nutrients occurs largely from the small intestine.

1. The absorptive surface area of the small intestine is increased greatly by folds, villi, and **microvilli**.

2. **Water** absorption by osmosis occurs secondarily to the absorption of dissolved minerals.

3. **Sodium** is actively transported across the brush border membrane into the intestinal cell from which it is extruded into blood by the sodium pump.

4. Other inorganic ions (e.g., calcium, potassium, zinc, chloride, phosphate, bicarbonate) and iron and iodide are actively or passively transported into the blood.

5. **Carbohydrates** are enzymatically digested into monosaccharide units, which may be absorbed by different mechanisms:
   a. Facilitated diffusion (fructose).
   b. Active transport by carrier-mediated process (glucose).
   c. Co-transport (glucose, galactose coupled with sodium).

6. **Amino acids** are transported by active processes.
   a. Each class of amino acid has its own specific carrier-mediated system in which amino acid transport is coupled with sodium transport.
   b. Dipeptides and tripeptides are also absorbed.
   c. In the newborns of some species, proteins are absorbed by pinocytosis.

7. **Triacylglycerols** are emulsified by bile salts and then digested by pancreatic lipase into a monoglyceride and free fatty acid mole cules. Bile salts render monoglycerides and free fatty acids water soluble by facilitating the formation of micelles.

8. **Monoglycerides and free fatty acids** are absorbed as individual molecules from the monomolecular dispersion.
   a. **Micelles** serve as a reservoir for these molecules and replenish them in the **monomolecular dispersion**.
   b. There is a dynamic relationship between the monomolecular dispersion and the micelles.
   c. Inside epithelial cells monoglycerides and free fatty acids are resynthesized to triacylglycerols, which aggregate as droplets.
   d. A layer of phospholipid and protein is added to each droplet to form a **chylomicron**, which enters the central **lacteal** of the villus and the lymphatic system.

VII. Processes of Absorption, Secretion, and Motility in the Large Intestine

A. Water and sodium are absorbed and potassium and bicarbonates are secreted by the large intestine.

1. The large intestine is subdivided into the cecum (and its appendix); ascending, transverse, and descending colon; sigmoid colon; rectum; and anus, with its internal and external **sphincters**.

2. The primary absorptive process is the active transport of sodium from the colon, which is accompanied by the osmotic absorption of water.

3. The colonic mucosa secretes an alkaline fluid that protects the mucosa from chemical and mechanical trauma and lubricates the fecal mass, thus facilitating its passage.

B. Motility of the large intestine is characterized by both mixing and propulsive movements.

1. Segmenting contractions promote absorption of water and inorganic ions by exposing the contents to the mucosa.

2. Propulsive mass movements may be stimulated by gastrin, the **gastrocolonic reflex**, and distention of colonic walls.

C. *The* ***defecation reflex****, when elicited by distention of the rectum, results in propulsion of feces to the outside of the body.*

1. Mass contractions move the contents into the rectum, whose distention elicits the defecation reflex.

2. **Constipation**, the accumulation of feces in the large intestine above the normal capacity, occurs when the muscular activity of the large intestine is decreased.

Key Terms

acini
bile (salts, pigments)
chief cells
cholecystokinin
chylomicron
defecation reflex
enterogastric inhibitory reflex
enterohepatic circulation of bile salts
esophagus
gastric inhibitory peptide
gastrin
gastrocolic reflex
lipase
long reflexes
lower esophageal sphincter
micelles
microvilli
pancreatitis
parietal cells
pepsin
peristalsis
portal circulation
short reflexes
sphincter
splanchnic circulation
zymogen granules

Testing Your Knowledge

I. Matching. Select the correct term from the Key Terms.

_________ 1. epithelial lining of digestive tract

_________ 2. path for blood from stomach and intestines to the liver

_________ 3. secreted by all regions of the digestive tract

_________ 4. reflex mediated through nerve cells lying entirely within digestive tract wall

_________ 5. reflex mediated over nerves outside of the tract

_________ 6. secreted by G cells, stimulates stomach and secretion and gastric emptying

_________ 7. hormone from small intestine, stimulates bicarbonate secretion from pancreas and biliary system

_________ 8. prevents reflux of gastric contents into esophagus

_________ 9. propulsive movement in esophagus

_________ 10. sphincter between stomach and duodenum

_________ 11. secrete pepsinogen

_________ 12. pancreatic exocrine secretory cells

_________ 13. act together to digest triacylglycerides

_________ 14. membranous projections that increase surface area

_________ 15. primary site of water absorption

II. Dichotomous. Circle the correct choice from the paired terms in parentheses.

1. The mesenteric artery is part of the (splanchnic/portal) circulation.

2. (Short/Long) reflexes are mediated by nerve cells lying entirely in the wall of the digestive tract.

3. Gastric emptying is stimulated by (gastrin/secretin).

4. Vagal afferent fibers and mechanoreceptors play roles in the (cephalic/gastric/intestinal) phase of gastric secretion of HCl.

5. Pepsinogen is secreted by (chief/parietal/acinar) cells in the (stomach/pancreas).

6. There is a (direct/inverse) relationship between the concentration of bile salts in the blood and volume of bile secreted.

7. Gastric emptying is retarded if the intestinal contents are (hyperosmotic/iso-osmotic/hypo-osmotic).

8. The (pyloric/ileocecal) sphincter controls the movement of the contents of the small intestine to the large intestine.

9. Monoglycerides and free fatty acids move through epithelial cells in a complex called a (micelle/chylomicron).

10. Propulsive mass movements in the large intestine are attributed to a long reflex called the (gastrocolic/defecation) reflex.

III. True/False. In the space to the left of the number write T for true or F for false.

_______ 1. The esophagus contains only smooth muscle.

_______ 2. If the stomach were lacking parietal cells, the activity of pepsin would be increased.

_______ 3. Vomiting may lead to disturbances in water volume, electrolyte balance, and acid-base balance.

_______ 4. Pancreatic enzymes require a pH around 7 for optimal activity.

_______ 5. Secretin stimulates the release of bile from hepatocytes.

_______ 6. The formation of gallstones occurs when the enterohepatic circulation of bile salts is defective.

_______ 7. During the absorption process, an osmotic gradient favors movement of water from blood to the lumen of the small intestine.

_______ 8. The unaltered lipid-soluble digestive products of triacylglycerides move easily from the lumen through epithelial cells of the brush border into the lymph.

_______ 9. A toddler is "potty trained" when he/she has learned to control his/her internal anal sphincter.

_______ 10. A deficiency of lactase may lead to an electrolyte imbalance.

IV. Sentence Completion. Write your response in the space provided.

1. The __________ nerve plays a major role in esophageal peristaltic movements.

2. The enterogastric inhibitory reflex occurs when __________ activity is decreased and autonomic nervous system __________ activity is increased.

3. HCl suppresses the release of __________, which stimulates production of HCl.

4. __________ is necessary for the secretion of HCl by parietal cells and pancreatic bicarbonate.

5. Trypsinogen is converted to __________ by the action of __________.

6. __________ is released from duodenal mucosa in response to gastric acid.

7. The __________ increase the surface area of the small intestine at least 600 times.

8. __________ and __________ are required for normal digestion and absorption of triacylglycerols.

9. The osmotic absorption of water from the intestine (large and small) depends on active transport of __________.

10. The external sphincter consists of __________ muscle and is under __________ control.

V. Multiple Choice. Circle the letter of the best response.

1. If the vagal somatic efferent fibers were cut, which of the following activities would be most affected?
   a. swallowing
   b. secretion of saliva
   c. chewing
   d. enterogastric inhibitory reflex
   e. vomiting

2. Which of the following pairs (exocrine site and product) is mismatched?
   a. salivary glands — amylase, mucin
   b. esophagus — mucin
   c. stomach — intrinsic factor

d. pancreas — bicarbonate
e. small intestine — amylase

3. The enterogastric inhibitory reflex may be initiated by all of the following except:
   a. gastric inhibitory peptide
   b. secretin
   c. gastrin
   d. a diet high in fat
   e. increase of sympathetic activity in stomach

4. All of the following sets (exocrine cell and product) are correctly matched except:
   a. parietal cell — intrinsic factor
   b. chief cell — pepsinogen
   c. acinar cell — amylase
   d. hepatocyte — bicarbonate
   e. parietal — HCl

5. The activity of pepsin in the gastric lumen depends directly or indirectly on all of the following except:
   a. gastrin
   b. carbonic anhydrase
   c. pepsin
   d. histamine
   e. secretin

6. Which of the following reactions occurs in the gastric secretion of HCl and pancreatic secretion of bicarbonate?
   a. exchange of HCO_3 for Cl^-
   b. exchange of H^+ for Na^+
   c. production of HCO_3^-
   d. secretion of H^+ into lumen
   e. secretion of $HCO3^-$ into lumen

7. Which of the following sets (enzyme — substrate — product) is mismatched?
   a. amylase — peptides — amino acids
   b. trypsin — proteins — small peptides
   c. lipase — triacylglycerols — monoglycerides
   d. peptidase — peptides — amino acids
   e. phospholipase — lecithin — lysolecithin

8. If the sodium pump is inhibited by a drug, which of the following processes would be directly affected?
   a. absorption of glucose across the brush border
   b. secretion of HCl and pancreatic bicarbonate
   c. recycling of bile salts
   d. absorption of amino acids across the brush border
   e. a and d are correct

9. The correct sequence for absorption and movement of the digestive products of triacylglycerols from the human to lymph is:
   a. formation of large oil droplets → emulsified oil droplets → micelles → monomolecular dispersion → chylomicrons
   b. emulsified oil droplets → large oil droplets → micelles → monomolecular dispersion → chylomicron
   c. formation of large oil droplets → emulsified oil droplets → chylomicrons → monomolecular dispersion → micelles
   d. formation of chylomicrons → emulsified oil droplets → micelles → monomolecular dispersion
   e. none of the above

10. Gastrin stimulates all the following except:
    a. gastrocolic reflex
    b. gastric emptying
    c. secretion of pancreatic bicarbonate
    d. gastric acid secretion
    e. mucosal growth of the stomach and small intestine

VI. Complete the Table.

For each nutrient, indicate the digestive enzyme, digestive product(s), and the mechanism by which the digestive product(s) is absorbed.

Nutrient/Source	Digestive Enzyme(s)	Digestive Product(s)	Mechanism of Absorption
Starch			
Cellulose			
Myosin			
Actin			
Cream			
Beef hamburger			
Roll			
Apple			
Casein			

VII. Diagram the Figure.

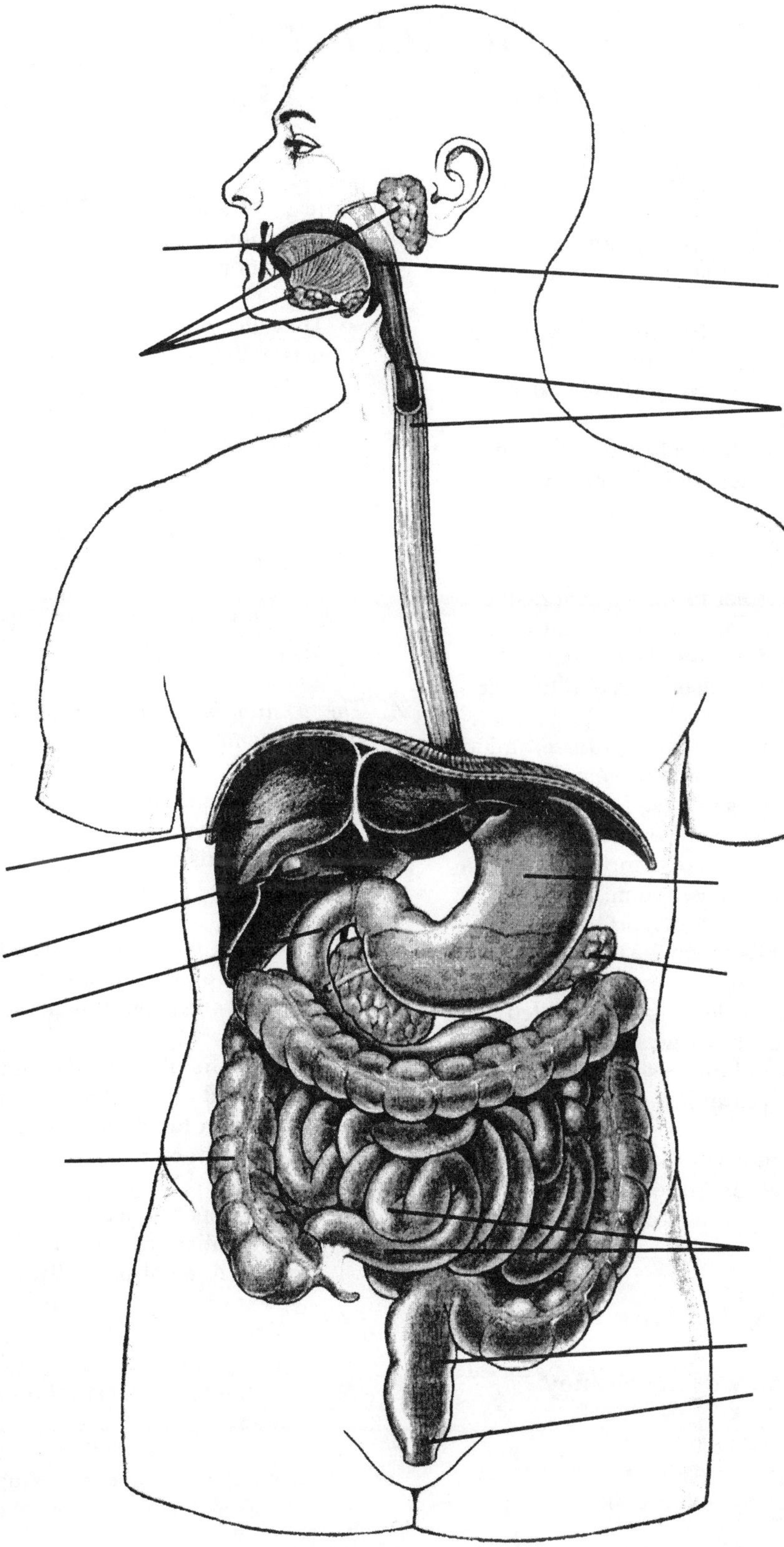

Chapter 23

The Kidney

Chapter Learning Objectives

1. Outline the functions of the kidney.
2. Diagram and label the gross anatomy of the kidney.
3. Diagram, label, and describe the function of a nephron and each of its parts.
4. Distinguish between cortical and juxtamedullary nephrons.
5. Describe the blood vessels that supply and drain the kidney and how the blood flow is regulated. Trace a drop of blood through the kidney from the renal artery to the renal vein.
6. Describe glomerular filtration, factors that influence it, and how glomerular filtration rate is measured. Define clearance.
7. Describe tubular reabsorption and tubular secretions.
8. Describe the reabsorption of glucose, amino acids, uric acid, urea, and proteins as to the transport processes involved for each type of molecule. Describe where each is reabsorbed and the effects of competition and concentration on reabsorption.
9. Describe the tubular transport of mineral electrolytes. (a) Describe the transport of sodium through the proximal convoluted tubule, loop of Henle, distal convoluted tubule, and collecting ducts. (b) Describe the conditions and processes by which potassium and phosphate are secreted and reabsorbed.
10. Explain how the loop of Henle operates.
11. List some of the special substances produced by the kidney and describe what each does.

Expanded Chapter Outline

I. Functional Anatomy of the Kidney

~ The kidneys regulate (a) osmolarity of body fluids; (b) the excretion of sodium and water; (c) volume of extracellular fluids; (d) concentration of individual electrolytes (e.g., K^+, Ca^{2+}, Mg^{2+}, Cl^-, sulfate and phosphate ions); (e) plasma HCO_3^- and H^+ concentration.

~ They also eliminate metabolic waste products, produce hormones (e.g., erythropoietin and thromboxane), and synthesize a variety of molecules (e.g., vitamin D, ammonia, and glucose).

~ The two parts of the kidney are (a) the medulla, in which the loops of Henle and medullary collecting ducts are located, and (b) the cortex, which contains the glomeruli, convoluted tubules, and cortical collecting ducts.

A. The nephron is the basic unit of kidney structure and function.

1. The basic unit of kidney structure and function is the **nephron**, which is composed of a **renal corpuscle** and its attached **renal tubule**.
2. The renal tubule consists of a tuft of capillaries, the **glomerulus**, surrounded by a double-walled cup, **Bowman's capsule**.
3. The renal tubule is divided into **proximal convoluted tubule, loop of Henle**, and **distal convoluted tubule**, which leads to **collecting ducts**.
4. The two types of nephrons are cortical and juxtamedullary nephrons.
   a. They differ structurally in the size of their glomeruli, length of limbs, and location in the kidney.
   b. They differ functionally in filtration rate, tubular transport properties, and renin content.
5. The renal tubule and collecting duct are composed of a single layer of epithelial cells

surrounding a tubule lumen. Cell structure and function vary considerably from one segment to another.

B. Blood vessels and nerves enter and leave the kidney at its hilum.

1. The renal artery from the abdominal aorta divides into smaller arteries that course through the medulla and cortex.The cortical radial arteries supply the **afferent arterioles,** each of which leads to a glomerulus.

2. The **efferent arterioles** from the glomeruli break up into the **peritubular capillaries,** which form an extensive network surrounding the tubules and lead to veins that eventually empty into the **renal vein.**

3. The glomeruli, in which blood pressure is high, are specialized for filtration, and the peritubular capillaries are specialized for uptake of fluid reabsorbed by the tubules.

4. Sympathetic efferent fibers predominantly innervate blood vessels and cause vasoconstriction.

5. Lymphatic vessels are found in the cortex.

C. The juxtaglomerular apparatus produces renin and controls the rate of glomerular filtration.

1. The **juxtaglomerular apparatus** is a complex of the macula densa, extraglomerular mesangial cells, and granular cells.

2. The close relationship may facilitate communication between blood vessels and tubules.

D. The kidneys have a high blood flow.

1. The high blood flow to the kidneys is necessary to sustain a high rate of **filtration** in the glomeruli.

2. Blood flow is not distributed uniformly within the kidney.

3. Renal **autoregulation** prevents changes of arterial blood pressure from producing large changes in glomerular filtration rate.

4. Renal blood flow is affected by hormones and physiological and physical conditions.

II. Processes Involved in Formation of Urine

A. Glomerular filtration involves ultrafiltration of plasma at a high rate.

1. Three basic processes are involved in urine formation: glomerular filtration, tubular reabsorption, and tubular secretion.

2. **Excretion** refers to what comes out in the urine that leaves the body and depends on glomerular filtration and tubular transport.

3. **Ultrafiltration** is a nonselective process in which a filtrate of plasma is pushed out of the glomerulus by high **hydrostatic pressure** into Bowman's capsule.

4. The glomerular filtration barrier consists of three layers: capillary endothelium, glomerular basement membrane, and **podocyte cell** layer.

5. The filtration barrier is a size-selective and electrostatic barrier to the free movement of molecules.
   a. Fenestrae in the endothelial layer and slit pores between podocytes provide openings for the passage of small molecules.
   b. Negatively charged glycoproteins on barrier elements impede the passage of negatively charged plasma proteins.

6. Factors that affect **glomerular filtration rate (GFR),** include:
   a. **Glomerular capillary hydrostatic pressure (P_{GC}),** the driving force for filtration.
   b. **Hydrostatic pressure** in the urinary space in Bowman's capsule (**P_{BS}**).
   c. **Intracapillary colloid osmotic pressure** caused by the plasma proteins (π_{GC}).

7. The net filtration pressure gradient is equal to the difference between pressures favoring and opposing filtration and is expressed by the equation:

$$\text{net filtration pressure} = P_{GC} - P_{BS} - \pi_{GC}$$

8. The glomerular filtration rate also depends on the glomerular filtration coefficient, Kf, and is also expressed as:

$$\text{GFR} = K_f \times \text{net filtration pressure}$$

9. **Glomerular filtration rate** is high because:
   a. The glomerular filtration barrier has a high fluid permeability.
   b. The glomerular hydrostatic pressure is unusually high for a capillary bed.
   c. The rate of renal blood flow is high.

10. Glomerular hydrostatic pressure depends mainly on arterial blood pressure and resistance of upstream blood vessels.

11. The measurement of GFR is important in assessing kidney function.

12. **Clearance** (C) is the rate of excretion of a substance (x) that is equal to the product of its urine concentration (U_x) and urine flow rate (V) divided by its plasma concentration (P_x) and is expressed by the equation:

$$C_x = U_x \times V/P_x$$

B. Tubular reabsorption returns filtered substances back to the blood.

1. In this selective process, substances are transferred out of the tubular fluid and returned to the peritubular capillaries.

2. Reabsorption may be either active or passive.

3. Glucose is actively reabsorbed by the proximal tubule.
   a. Glucose is cotransported with sodium across the luminal cell membrane.
   b. The maximal rate of glucose reabsorption is called the **tubular transport maximum (Tm)** for glucose (G).
   c. Other sugars are reabsorbed by the same mechanisms as glucose.
   d. Sugars may compete for transport.

4. There are several distinct transport mechanisms for amino acids, each reabsorbing a group of structurally similar compounds.
   a. Amino acid reabsorption is active and depends on sodium transport.
   b. It occurs in the proximal tubule and shows a maximal rate of transport and within-group competition.

5. Reabsorption and secretion of uric acid occur in the proximal tubule by carrier-mediated processes.

6. Eighty percent of urea, the major nitrogen-containing end product of metabolism, is reabsorbed by the inner medullary collecting duct, and 50% is recycled by the loops of Henle.

7. The proximal tubules reabsorb a variety of proteins by endocytosis.

8. Many other organic compounds are reabsorbed by the kidney tubules.

C. Proximal tubules actively secrete many organic anions and organic cations.

1. In this process, substances move from the peritubular capillaries into the tubular fluid.

2. Secretory mechanisms that transport a variety of organic anions and cations are present in the proximal tubule.
   a. They also show maximal transport rate at high plasma levels.
   b. They also show within-group competition for transport.

III. Tubular Transport of Mineral Electrolytes

A. Sodium reabsorption is the major activity of the kidney tubules.

1. It is the major driving force for water reabsorption.

2. It is coupled with the reabsorption of both organic and inorganic substances and is related to the secretion of hydrogen and potassium ions.

3. The proximal convoluted tubule absorbs 70% of the filtered sodium and water.

4. Sodium moves into the cell along chemical and electrical gradients.
   a. Its entry is coupled with the movement of other solutes, such as glucose, amino acids, phosphate ions, and chloride ions.
   b. The intracellular concentration is low because sodium is actively transported out of the basolateral membrane.

5. Reabsorbed fluid is taken into the peritubular capillaries by a difference in hydrostatic and colloid osmotic pressures.

6. The **loop of Henle** absorbs about 20% of filtered sodium and 10% of the filtered water.

a. The **ascending limb** is impermeable to water but reabsorbs salt by a coupled sodium-potassium-chloride transporter in the terminal cell membrane and a sodium-potassium pump in the basolateral cell membrane.
b. Most of the remaining salt and water is resorbed in the distal collecting tubule and collecting duct by active transport for sodium.

7. The collecting ducts are the site of fine control for sodium and water reabsorption by the hormones aldosterone and **ADH**, respectively.

B. Potassium is filtered, reabsorbed, and secreted in the kidneys.

1. Potassium is reabsorbed primarily in the proximal convoluted tubule and loop of Henle. It is secreted by the collecting ducts under normal conditions or with potassium excess.

2. Potassium is taken up into the cortical collecting duct principal cell by the basolateral membrane sodium-potassium pump and then diffuses across the luminal cell membrane into the lumen.

3. The concentration of potassium in the cell is affected by aldosterone and the extracellular potassium concentration. Aldosterone increases luminal membrane permeability to sodium and potassium.

4. Across many nephron segments is a transtubular potential difference, which is less negative in the luminal side than at the basolateral side of the cell. This favors potassium secretion into the lumen.

C. Calcium, magnesium, and phosphate are reabsorbed by the kidney tubules.

1. Renal excretion of calcium represents only about 10% of total daily output from the body. The rest passes out of the body in the feces.

2. About 60% of filtered calcium is reabsorbed in the proximal convoluted tubule and most of the rest in the thick ascending limb, distal convoluted tubule, and collecting ducts.

3. About 20–30% filtered magnesium is reabsorbed in the proximal convoluted tubule and 50-60% reabsorbed in the loop of Henle.
   a. Only about 3–5% is excreted in the urine, and about two-thirds of dietary magnesium is lost in the feces.
   b. The kidneys play a major role in regulating plasma magnesium concentration.

4. The kidneys play an important role in regulating the plasma concentration of inorganic phosphate. Filtered phosphate is reabsorbed by sodium-dependent, secondary active transport in the proximal tubule.

5. Phosphate excretion depends on the Tm for phosphate, which is normally close to the filtered phosphate load.
   a. Tm will be exceeded more than usual, and phosphate excretion will increase.
   b. If filtered phosphate amounts are decreased, all will be reabsorbed.

IV. Tubular Reabsorption of Water

A. The kidneys save water when they produce a hyperosmotic urine.

1. The rate of water excretion in the urine depends on the glomerular filtration and tubular reabsorption rates.

2. The ability to produce a hyperosmotic urine or concentrated urine depends on the length of the loop of Henle: the longer the loop, the greater the ability to produce hyperosmotic urine.

B. The countercurrent hypothesis explains the formation of osmotically concentrated urine.

1. The loops of Henle act as **countercurrent multipliers** in setting up an increasing gradient of osmolality from the junction of the cortex and medulla to the tip of the renal papillae.

2. The blood vessels of the medulla acting as passive countercurrent exchangers help preserve the gradient.

3. The collecting ducts acting as osmotic equilibrating devices alter the osmolality of the urine, depending on the plasma level of ADH.

4. Countercurrent multiplication is an energy-consuming process.

5. The osmotic gradient along the loop of Henle can be established by salt transport between ascending and descending limbs.

6. The osmolality of the tubular fluid changes as the fluid passes through the descending and ascending limbs and the collecting ducts in the outer and inner medulla.

C. Urea plays a special role in the concentrating mechanism in the inner medulla.

1. **Urea** is added to the inner medulla from the collecting ducts and from the calyceal urine bathing the papilla.

2. The osmotic pressure caused by urea sets up a concentration gradient for sodium in the inner medulla.

D. The vasa recta are countercurrent exchanges.

1. The **vasa recta** are responsible for removing water from the medulla by a plasma colloid osmotic pressure mechanism.

2. Water enters the medullary interstitium from the descending limb of the loop of Henle and collecting ducts.

E. Many factors influence the ability to form an osmotically concentrated urine.

1. Factors affecting urinary concentrating ability include:
   a. Antidiuretic hormone (ADH).
   b. Adequate delivery to and reabsorption of salt along the ascending limb.
   c. Adequate delivery of fluid to the collecting duct.
   d. Adequate urea concentration.

2. The length of the loop of Henle is also a factor.

F. Excretion of an osmotically dilute urine rids the body of excess water.

1. Production of an osmotically dilute urine results from the reabsorption of solute across a relative water-impermeable membrane, with water left behind.

2. The degree of water permeability of the collecting ducts depends on the presence of ADH.

V. Functions of the Ureters and Urinary Bladder

A. The ureters transmit the urine to the urinary bladder.

1. Urine moves through the ureters by peristaltic movements in the muscle.

2. The **ureters** are innervated by sympathetic and parasympathetic nerve fiber.

*B. The **urinary bladder** stores the urine and is periodically emptied.*

1. Smooth muscle is found in the bladder and internal sphincter; skeletal muscle is found in the external sphincter.

2. The bladder and its neck are innervated by parasympathetic pelvic nerves and sympathetic hypogastric nerves; the external sphincter is innervated by somatic pudendal nerves.

3. The two functions of the bladder are storage and excretion.

*C. **Micturition** is a complex act.*

1. The basic reflex occurs at the level of the sacral spinal cord but may be modified by messages from centers in the midbrain and cerebral cortex.

2. Stimulation of stretch bladders triggers the reflex, i.e., contraction of bladder and relaxation of sphincters.

3. The reflex is released by inhibitory impulses from the cerebral cortex.

Key Terms

aldosterone
antidiuretic hormone (ADH)
Bowman's capsule
clearance
colloid osmotic pressure secretion
countercurrent exchanger
countercurrent multiplier
distal convoluted tubule
excretion
fenestra
filtration coefficient (K_f)
glomerular filtration
glomerular filtration rate (GFR)
glomerular hydrostatic pressure
glomerulus
hydrostatic pressure
juxtaglomerular apparatus
juxtamedullary nephron
loop of Henle
luminal membrane
nephron
peritubular capillary
podocyte
pressure (P_{GC})
proteinuria
proximal convoluted tubule
renal plasma
renal tubule
tubular reabsorption
tubular transport maximum (T_m)
urea
ureter
vasa recta

Testing Your Knowledge

I. Matching. Select the correct term from the Key Terms.

________ 1. tuft of capillaries in renal corpuscle

________ 2. major site of reabsorption of glucose, amino acids

________ 3. tube draining kidney

________ 4. have large glomeruli and long loops of Henle

________ 5. membrane through which solutes in tubular fluid pass

________ 6. surround tubules and specialized for uptake of fluids reabsorbed by tubules

________ 7. may play a role in regulating glomerular blood flow and filtration rate at level of single nephron

________ 8. forces fluid from renal corpuscle into renal tubule

________ 9. hole in endothelial cell layer lining capillaries

________ 10. exceeded for glucose in diabetes mellitus

________ 11. function as countercurrent multipliers

________ 12. passive countercurrent exchangers in medulla

________ 13. increases with increase in concentration of serum proteins

________ 14. affects water permeability of collecting tubules

________ 15. primary nitrogenous waste product

II. Dichotomous. Circle the correct choice from the paired terms in parentheses.

1. Large glomeruli, long thin limbs, and long loops are characteristic of (cortical/juxtamedullary) nephrons.

2. You are studying an electron micrograph (EM) of cells from a proximal tubule. You identify several mitochondria on the left, numerous microvilli, and a tight junction on the right side of the EM. You also expect to see the lumen on the (right/left) and a basement membrane on the (right/left).

3. If blood pressure is increased, the cortical radial arteries and afferent arterioles (constrict/dilate), thereby (increasing/decreasing) renal vascular resistance, and (increasing/decreasing) blood flow.

4. Consider two kidneys, A and B, with the same PGC. If PBS in A is less than PBS in B, the net filtration of A will be (greater/less) than the net filtration in B. The glomerular filtration of A will be (greater/less) than B.

5. A(n) (increase/decrease) value of plasma creatinine relative to a normal plasma value would indicate an impaired GFR.

6. Many proteins are reabsorbed by (endocytosis/exocytosis) in the (distal/proximal) tubules.

7. You would expect to see more mitochondria at the (luminal/basolateral) surface of a proximal tubule cell than at the (luminal/basolateral) surface.

III. True/False. In the space to the left of the number write T for true or F for false.

_______ 1. The lumen is located next to the brush border of the basolateral membrane in tubule cells.

_______ 2. The perfusion rate of blood through the kidney is less than that through the brain, liver, or heart.

_______ 3. Dietary intake of calcium and sodium has no effect on the amount excreted.

_______ 4. PGC is always greater than PBS for a normal kidney.

_______ 5. Kf is a constant and is not affected by physiological changes in the body.

_______ 6. Glucose is co-transported with sodium across the basolateral membrane of the tubule cell into the peritubular capillaries.

_______ 7. Penicillin is a good drug for treating urinary tract infections because it is secreted by kidney tubules.

_______ 8. Colloid osmotic pressure is the force driving the uptake of water into peritubular capillaries.

_______ 7. A person lacking the posterior pituitary would produce a very hypo-osmotic urine.

_______10. A person taking a "loop" diuretic would most likely produce more volume and a saltier urine than a healthy person.

IV. Sentence Completion. Write your response in the space provided.

1. Glomeruli and cup-shaped capsules are found in the __________ part of the kidney.

2. If for substance X, amount excreted is greater than amount filtered, X is ________.

3. You would not expect to find __________ in the urine of a healthy person.

4. If 270 mmol of calcium is filtered and 266 mmol is reabsorbed, ____________ mmol would be excreted under normal conditions.

5. If P_{GC} is 60 mm Hg, π_{GC} = 30 mm Hg and P_{BS} = 12 mm Hg, the net filtration rate is __________.

6. The reabsorption of both simple sugars and amino acids depends on __________.

7. You know the __________ and __________ have been exceeded if glucose is measured in the urine.

8. Active absorption of sodium occurs in the __________ convoluted tubule.

9. Both aldosterone and ADH act on the __________ to increase reabsorption of __________ and __________, respectively.

10. Urine flows from the kidney through the __________ to the __________ for storage and through the __________ during urination.

V. Multiple Choice. Circle the letter of the best response.

1. Blood flows from afferent arterioles into:
   a. glomeruli
   b. Bowman's capsules
   c. peritubular capillaries
   d. cortical radial arterioles
   e. efferent arterioles

2. Total renal blood flow is decreased under all the following conditions except:
   a. strenuous exercise
   b. epinephrine
   c. prostaglandins E_2 and I_2
   d. angiotensin II
   e. cold

3. For which molecules is the equation: secreted = excreted – filtered > zero?
   a. sodium
   b. bicarbonate
   c. creatinine
   d. urea
   e. uric acid

4. Under normal conditions, glomerular filtration rate can be decreased by a(n):
   a. increase in K_f
   b. increase in P_{BS}
   c. increase in P_{GC}
   d. decrease in π_{GC}
   e. b and d are correct

5. The glomerular filtration rate of a substance, Z, would equal clearance rate of Z if Z is:
   a. $filtered_z = excreted_z + secreted_z$
   b. $filtered_z > excreted_z$
   c. $filtered_z < excreted_z$
   d. $filtered_z = excreted_z$
   e. $filtered_z - secreted_z = excreted_z$

6. Calculate GFR for insulin (numerical values do not reflect real values): PIN = 0.35 mg/ml and UIN = 25 mg/ml. 900 ml of urine is collected in a 12-hour period.
   a. 121 ml plasma/min
   b. 89 ml plasma/min
   c. 31 ml plasma/min
   d. 9 ml plasma/min
   e. 45 ml plasma/min

7. All of the following transport activities occur at the luminal membrane except:
   a. Na^+/H^+ exchange
   b. Na^+/sugar cotransport
   c. transport of Na^+ against its concentration gradient
   d. Cl^-/HCO_2^- exchange
   e. Na^+/amino acid cotransport

8. Potassium secretion is favored by:
   a. aldosterone
   b. a less negative electrical potential at the luminal surface than at the basolateral surface
   c. a high intracellular K^+
   d. an increased transtubular potential difference
   e. all of the above

9. As a result of countercurrent exchange, the fluid in the tubule is most hyperosmotic in the:
   a. proximal convoluted tubule
   b. distal convoluted tubule
   c. descending limb of loop of Henle
   d. ascending limb of loop of Henle
   e. loop of Henle in inner medulla

10. Under which condition would hyperosmotic urine be produced?
    a. neurogenic diabetic insipidus
    b. nephrogenic diabetic insipidus
    c. low-protein diet
    d. short loop of Henle
    e. insufficient medullary flow

VI. Complete the Table.

For each condition, indicate the effect (increase, decrease, none) on glomerular filtration rate.

Condition	Glomerular Filtration Rate
1. increase in length of loop of Henle	
2. decreased glomerular hydrostatic pressure	
3. decrease in renal blood flow	
4. concentration of glucose above T_m	
5. decrease in sodium reabsorption	
6. constriction of afferent arterioles	
7. increase in hydrostatic pressure in Bowman's space	
8. decrease in plasma colloid osmotic pressure	

VII. Complete the Table.

For each condition, indicate the direct effect (none, increase, decrease) on tubular reabsorption of glucose, sodium, or water.

Condition	Glomerular Filtration Rate		
	Glucose	Sodium	Water
1. high plasma concentration of fructose			
2. low plasma concentration of sodium			
3. presence of aldosterone			
4. presence of antidiuretic hormone			
5. diabetes insipidus			
6. deficiency of dietary protein			
7. decrease in length of loop of Henle			

Chapter 24

Regulation of Fluid and Electrolyte Balance

Chapter Learning Objectives

1. Describe the electrolyte composition and balance of body fluids.
2. Describe factors affecting the distribution of water between cellular and extracellular fluids and between blood plasma and interstitial fluid.
3. Describe mechanisms that regulate water balance, and describe body responses to cellular dehydration and extracellular dehydration.
4. Indicate how sodium is balanced in the body. Describe body responses to changes in dietary intake of sodium.
5. Explain the role of hormones on fluid balance.
6. Consider the regulation of potassium balance in the body. Discuss factors affecting the distribution of potassium between cells and extracellular fluid and potassium excretion.
7. List disturbances occurring in response to uremia.
8. Explain the different kinds of kidney dialysis.

Expanded Chapter Outline

I. Body Fluids and Fluid Compartments

A. Body water is distributed in several fluid compartments.

1. The percentage of total body weight that is water varies from 45% to 75% and is related to the amount of body fat.

2. Two-thirds of body water (40% of body weight) is contained in the intracellular fluid compartment.

3. One-third of body water is in the extracellular fluid compartment, which is subdivided into:
   a. The blood plasma (water = 5% body weight).
   b. Interstitial fluid and lymph (water = 15% body weight).
   c. Transcellular fluid subcompartments, e.g., cerebrospinal fluid, aqueous humor (1% of body weight).

B. Body fluids differ in electrolyte composition.

1. Electrolytes contribute most to the osmotic pressure of the body fluids and are crucial in the distribution of body water.

2. Concentrations of electrolytes are expressed in terms of **milliequivalents** per liter solution.

3. Sodium is the major cation; chloride and bicarbonate ions are the major anions in the plasma.

4. The differences in concentrations of small ions between the plasma water and interstitial fluid arise because of differences in protein concentration between these two fluids.

5. The distribution of potassium and sodium ions is maintained by the sodium-potassium pump, whereas that of chloride and bicar bonate is accounted for by the membrane potential difference.

6. The intracellular and organelle distribution of ions is not uniform.

C. The osmolalities of intracellular and extracellular fluids are normally identical.

1. **Osmotic pressure** is of prime importance in determining the distribution of water between cells and extracellular fluid and is directly related to **osmolality**.

2. In extracellular fluid, the amount of sodium is a major determinant of the volume of water present. Sodium, chloride, and bicarbonate are the osmotically active solutes in the extracellular fluid.

3. In the intracellular fluid compartment,

potassium is the major osmotically active solute.

4. The sodium-potassium pump plays a key role in regulation of cell volume.

5. The normal distribution of water between the cellular and extracellular compartments changes in a variety of circumstances.

D. The distribution of water between blood plasma and interstitial fluid is determined by Starling forces.

1. The relative volumes of plasma and interstitial fluids are determined by the balance between capillary and tissue hydrostatic and collard osmotic pressures.

2. An increase in capillary hydrostatic pressure favors net movement of fluid out of the vascular system, while a decrease would favor movement of interstitial fluid into the circulation.

II. The Concept of Fluid and Electrolyte Balance

A. Water input and output are normally equal.

1. The composition of body fluids remains constant despite varying dietary intake of food and water.

2. The kidneys play a major role in maintaining balance with respect to water, mineral electrolytes, hydrogen ions, and various organic compounds.

3. The regulation of body water balance involves two basic mechanisms:
   a. The thirst sensation.
   b. Control of renal water excretion by the **antidiuretic hormone (ADH)**.

4. **Thirst** is controlled by centers in the hypothalamus that relay information to the cerebral cortex, where thirst becomes a conscious sensation. The hypothalamus controls the release of ADH.

5. Cellular dehydration is detected by **osmoreceptors** in the anterior hypothalamus, which signals the cerebral cortex to give rise to the thirst sensation.

6. Extracellular dehydration, or a decrease in effective circulatory blood volume, is detected by various **baroreceptors**, such as arterial baroreceptors in the carotid sinus and aortic arch and **stretch receptors** in large veins in the thorax. Impulses are sent to the **thirst control centers** in the hypothalamus.

7. ADH facilitates water reabsorption by the collecting ducts. Cellular and extracellular dehydration stimulate the release of ADH.

8. Renal excretion of water is continuously adjusted by changes in plasma ADH levels so that plasma osmolality and extracellular fluid volume are kept at normal levels.

B. The kidneys are the major site of sodium balance.

1. The kidneys are the major site of control of sodium output and normally adjust excretion so as to maintain balance. Sodium balance is related to regulation of extracellular fluid volume.

2. Changes in dietary input lead to appropriate changes in sodium excretion by the kidneys.

3. Effector mechanisms include changes in:
   a. Glomerular filtration rate.
   b. Plasma aldosterone levels.
   c. Peritubular capillary Starling forces.
   d. Renal sympathetic nerve activity.
   e. Intrarenal blood flow distribution.
   f. Plasma **atrial natriuretic factor**.

4. **Aldosterone** increases sodium reabsorption by a variety of epithelial tissues and the collecting ducts.

5. The **renin-angiotensin** system controls aldosterone secretion.
   a. Volume depletion stimulates the release of renin from granular cells of the juxtaglomerular apparatus.
   b. **Renin** acts on the plasma protein **angiotensinogen** to produce angiotensin I, which is converted enzymatically to angiotensin II.
   c. **Angiotensin II** raises blood pressure, stimulates the synthesis and release of aldosterone, and stimulates the hypothalamic thirst centers and the release of ADH.

6. Changes in peritubular capillary hydrostatic forces and colloid osmotic pressure affect the net reabsorption of salt and water by the proximal convoluted tubule.

7. Stimulation of renal sympathetic nerves results in a decrease in renal sodium excretion.

8. Atrial natriuretic factor is released by the atria in response to an increase in blood volume. It increases sodium excretion by increasing glomerular filtration rate and inhibiting sodium reabsorption.

9. Effective arterial blood volume is believed to be the regulated variable affecting sodium excretion. Volume receptors are located in the wall of the left atrium, in intrathoracic veins, and in the kidneys.

10. Changes in effective arterial blood volume can elicit the appropriate renal response by three possible mechanisms.
    a. A change in blood volume affects changes in glomerular blood flow and capillary pressure, which affect glomerular filtration rate.
    b. A change in blood volume is detected by an intrarenal baroreceptor, which then affects release of renin.
    c. A change in blood volume could change peritubular capillary Starling forces.

11. Other factors affecting sodium excretion include:
    a. Glucocorticoids.
    b. Estrogen.
    c. Osmotic diuretics.
    d. Poorly reabsorbed anions.
    e. Diuretic drugs.

C. *The kidneys are the major site of control of potassium balance.*

1. Potassium plays a number of important roles in the body.
    a. It influences the electrical excitability of cells and is the major osmotically active solute in cells.
    b. Potassium also affects acid-base balance and affects cell metabolism.

2. The kidneys are the major site of potassium loss and control of potassium balance.

3. Factors affecting the distribution of potassium between cells and extracellular fluid include:
    a. Activity of the sodium-potassium pump.
    b. Acid-base status of body fluids.
    c. Availability of insulin.
    d. Cellular breakdown due to trauma, infection, ischemia, and heavy exercise.

4. The regulation of plasma potassium involves hormones such as insulin, epinephrine, and aldosterone, and involves the kidneys.

5. Factors affecting potassium excretion include:
    a. Intracellular potassium concentration.
    b. Aldosterone.
    c. Excretion of anions.
    d. Urine flow rate.

III. Kidney Failure and Its Treatment

A. *Renal failure results in disturbances in many body systems.*

1. Kidney disease results from a loss of nephrons.

2. Disturbances occurring in conjunction with uremia, a symptom of renal failure, include:
    a. Anemia resulting from a deficiency of erythropoietin.
    b. Azotemia.
    c. Alterations in calcium and phosphate metabolism.
    d. Hypertension.
    e. Metabolic acidosis.
    f. Altered potassium and sodium balance.
    g. Impairment of urinary concentrating ability.

B. *Dialysis and transplantation are used to treat patients with terminal renal failure.*

1. Most of the signs and symptoms of uremia can be relieved by dialysis.

2. Two methods of dialysis may be used: peritoneal dialysis and hemodialysis, which uses an artificial kidney.

3. A major problem with kidney transplantation is immunological rejection by the host.

Key Terms

aldosterone
angiotensin (I, II)
angiotensin converting enzyme
antidiuretic hormone (ADH)
atrial natriuretic factor
azotemia
baroreceptor
dialysis (peritoneal)
diuretic hypovolemia
effective arterial blood volume
effective osmotic pressure
glomerulotubular balance
hemodialysis
hyperkalemia
hypertension
hypokalemia
hyponatremia
milliequivalent (mEq)
negative balance
osmoreceptor
osmotic diuretic
positive balance
renin
sodium appetite
water diuresis

Testing Your Knowledge

I. Matching. Select the correct term from the Key Terms.

________ 1. occurs when input exceeds output

________ 2. detects cellular dehydration

________ 3. detects hypovolemia

________ 4. low sodium concentration in the blood

________ 5. increases the water permeability of the collecting ducts

________ 6. equal to moles times valence of the ion

________ 7. increases reabsorption of sodium by collecting ducts

________ 8. increases sodium excretion by increasing glomerular filtration rate

________ 9. increase excretion of water

________ 10. below normal level of plasma potassium

________ 11. high blood pressure

________ 12. accumulation of nitrogenous waste products

________ 13. uses artificial kidney

________ 14. smooth muscle in urinary bladder wall

II. Dichotomous. Circle the correct choice from the paired terms in parentheses.

1. The (intracellular/extracellular) concentration of sodium is greater than the (intracellular/extracellular) concentration.

2. In extracellular fluid, the major determinant of water volume is (potassium/sodium), but in intracellular fluid the major osmotically active particle is (potassium/sodium).

3. In a positive balance, (input>output/output>input), but in a negative balance the opposite is true.

4. When plasma osmolality decreases, plasma ADH level (increases/decreases); when plasma osmolality increases, plasma level of ADH (increases/decreases).

5. ADH acts on the (proximal convoluted tubule/distal convoluted tubule/collecting duct).

6. A positive sodium balance results in a(n) (increase/decrease) of ECF volume.

7. Renin is released in response to a (decrease/increase) in ECF or effective arterial blood volume.

8. A(n) (increase/decrease) in peritubular capillary hydrostatic pressure and a(n) (increase/decrease) in colloid osmotic pressure reduce fluid uptake into the capillaries.

9. A plasma potassium concentration of 3.0 mmol would be described clinically as (hypokalemia/hyperkalemia).

10. A physiological salty solution is introduced into the abdominal cavity in (peritoneal/hemo-) dialysis.

III. True/False. In the space to the left of the number write T for true or F for false.

________ 1. The concentration of an electrolyte is the same regardless of body fluid compartment.

________ 2. If the Na^+/K^+-ATPase pump is

inhibited by lack of oxygen, the cell will swell.

_______ 3. A dry mouth is essential in regulating water intake.

_______ 4. A loss of 10% of blood volume is sufficient to trigger the thirst sensation.

_______ 5. Under a condition of total fasting, a person would produce hyperosmotic urine.

_______ 6. Destruction of the adrenal medulla would directly affect sodium reabsorption by the distal tubule.

_______ 7. Atrial natriuretic peptide is released by the atria in response to an increase in blood volume.

_______ 8. The regulated variable believed to affect sodium excretion is the extracellular fluid volume.

_______ 9. A person with Addison's disease may develop hypokalemia and should eat bananas as a preventive measure.

_______ 10. The blood plasma from a person with renal failure would show elevated levels of creatinine, urea, calcium, and sodium.

IV. Sentence Completion. Write your response in the space provided.

1. The plasma concentration of calcium is 5 mEq/liter. This is equal to ___________ mmol/liter.

2. Consider two aqueous solutions, A and B. The concentration of magnesium in A is 40 mEq/liter, while that in B is 40 mmol. If the solutions are separated by a H_2O permeable membrane, water will move from ________ to ___________. The osmolality of A is ___________ than that of B.

3. For a 50-kg female, the intracellular water is liters; the interstitial-lymph water is ______ liters; the plasma water is _________ liters; total body water is _________ liters; and extracellular fluid water is _________ liters.

4. Thirst is controlled by centers in the ________, which also controls the release of ___________.

5. Receptors for hypovolemia lie in the ______ and ________ system.

6. Angiotensinogen is converted to angiotensin I by _________, which is released by the _________ cells of the _________.

7. Inhibition of renal sympathetic nerve fibers promotes sodium _________.

8. The concentration of _________ is higher in intracellular fluid, whereas the concentration of ___________ is higher in extracellular fluid because Na^+/K^+ ATPase pumps ______ into the cell and _________ out of the cell.

9. Aldosterone favors _________ secretion by tubule cells but _________ reabsorption.

10. A _________ glomerular filtration rate may result in hyperkalemia.

V. Multiple Choice. Circle the letter of the best response.

1. The osmolality of the extracellular fluid is regulated in part by the:
   a. Na^+/K^+ ATPase activity
   b. concentration of sodium
   c. ADH
   d. concentration of potassium
   e. none of the above

2. The osmolality of extracellular fluid is 300 mosm/kg H_2O. Sodium chloride is added to the extracellular fluid. Which of the following describes what happens next?
   a. cellular osmolality <300; increase cellular volume
   b. cellular osmolality >300; decrease extracellular fluid volume
   c. extracellular fluid osmolality >300; increase cellular volume
   d. cellular osmolality >300; increase extracellular fluid volume
   e. no change in cellular fluid, osmolality or volume; increase extracellular fluid volume

3. The net movement of fluid from the plasma to interstitial fluid would be favored under

a condition of:
a. increase in plasma colloid osmotic pressure
b. increase in capillary hydrostatic pressure
c. decrease in capillary membrane permeability
d. a and b are correct
e. a and c are correct

4. The thirst sensation is triggered when (the):
a. osmorecepters in the anterior hypothalamus swell
b. effective plasma osmostic presssure is increased
c. plasma colloid osmotic pressure is decreased
d. hypervolemia occurs
e. effective arterial blood volume is increased

5. ADH release occurs under a condition of:
a. hypovolemia
b. decrease in effective arterial blood volume
c. stimulation of the barorecepters in the carotid sinus
d. hemorrhage
e. all of the above

6. You have ingested 2 liters of diet soda in about 15 minutes. Your body's immediate response is a(n):
a. decrease in water reabsorption
b. increase in water excretion
c. increase in ECF volume
d. decrease in plasma osmolality
e. decrease in release ADH

7. Angiotensin stimulates all the following glands or central nervous system components in response to a decrease in ECF except the:
a. hypothalamus
b. adrenal cortex
c. adrenal medulla
d. posterior pituitary
e. proximal tubule

8. Reduction in renal sodium excretion may result from all the following except:
a. decrease in GFR
b. increase in angiotensin II
c. increase in aldosterone
d. decrease in peritubular capillary forces
e. decrease in renal sympathetic nerve activity

9. The distribution of potassium between the intracellular and extracellular fluids is affected by all the following except:
a. Na^+/K^+ ATPase
b. plasma pH
c. presence of insulin
d. tissue trauma
e. all of the above affect the distribution

10. In response to a positive potassium balance, all the following occur except:
a. release of aldosterone
b. increase in permeability basolateral membrane of principal cells
c. increase in secretion of K^+
d. increase in activity of Na^+/K^+ ATPase
e. none of the above

VI. Complete the Table.

For each condition, indicate the effect (stimulate, inhibit, none) on water intake and release of antidiuretic hormones.

Condition	Water Intake	ADH Release
1. Ingestion of excess salt		
2. Decreased function of adrenal cortex		
3. Excessive sweating		
4. Activation of thirst sensation		
5. Decrease in plasma osmolality		
6. Cellular dehydration		
7. Hypovolemia		
8. Presence of renin in the plasma		
9. Presence of ADH		
10. Decrease in effective arterial blood volume		

VII. Complete the Table.

For each condition, indicate the effect (increase, decrease, none) on sodium reabsorption.

Condition	Sodium Reabsorption
1. Presence of ADH	
2. Heavy excercise	
3. Diarrhea	
4. Congestive heart failure	
5. Ingestion of salty potato chips	
6. Increase in glomerular filtration rate	
7. Hemorrage	
8. Presence of aldosterone	
9. Stimulation of renal sympathetic nerves	
10. Increase in peritubular capillary hydrostatic pressure	
11. Decrease in colloid osmotic pressure	
12. Presence of atrial natriuretic factor	
13. Osmotic diuretic	
14. Hyperkalemia	

Chapter 25

Regulation of Acid-Base Balance

Chapter Learning Objectives

1. Define acid and base, and explain the pH scale.
2. Explain how buffer systems work and give examples. Define pKa and pH, and write the Henderson-Hasselbalch equation.
3. Indicate the sources for hydrogen ions.
4. Consider the buffering systems of the body. List three chemical buffers, and describe how each operates as a buffer.
5. Describe how and under what conditions the respiratory system acts as a physiological buffer.
6. Indicate how the kidney participates in maintaining body pH.
7. Explain how the pH balance of the body can be upset.
8. Compare and contrast acidosis and alkalosis as to cause, physiological characteristics, and mechanisms operating in the body to minimize deviations from normal pH.
9. Distinguish clinically between metabolic and respiratory acidosis and alkalosis.

Expanded Chapter Outline

I. Principles of Acid-Base Physiology

A. *An acid is a proton donor, and a base is a proton acceptor.*

1. In a neutralization reaction, an acid reacts with a base to form a salt and water.

2. The amount of acid or base is expressed in terms of equivalents or **milliequivalents (mEq)**; one equivalent of acid neutralizes one equivalent of base.

B. *Amphoteric substances can function as acids and bases.*

1. Amino acids and proteins are **amphoteric** substances.

2. Proteins have many different side chain groups that can act as acids or bases.

C. *The acid dissociation constant is directly related to the strength of an acid.*

1. When an acid (HA) is added to water, it dissociates:

$$\mathbf{HA \Leftrightarrow H^+ + A^-}$$

2. The **rate of dissociation** ($HA \rightarrow H^+ + A^-$) is equal to the concentration of HA and the dissociation rate constant k_1.

3. The **rate of association** ($A^- + H^+ \rightarrow HA$) is equal to the product of the concentrations of H^+ and A^- and the association rate con stant k_2.

4. At equilibrium $k_1 \times [HA] = k_2 [H^+] \times [A^-]$ and with rearranging $[H+] \times [A-]/[HA] = k_1/k_2$. Thus, $\mathbf{k_1/k_2 = k_a}$. K_a is the equilibrium constant for the reaction or the dissociation or ionization constant for the acid.
   a. The higher the acid dissociation constant, the more completely an acid is dissociated or ionized.
   b. Strong acids have high dissociation constants.

5. **Acid dissociation constants** are often expressed in algorithmic form. $\mathbf{pK_a}$ is defined as the negative logarithm to the base 10 of K_a or $pK_a = -\log ka$, which may also be written as $pK_a = 10\ g\ (1/K_a)$

D. *The pH values of aqueous solutions range widely.*

1. pH is defined as the negative logarithm to the base 10 of the molar activity (concentration) of hydrogen ions or:

$$\mathbf{pH = -log\ (H^+) = log\ (1/(H^+))}$$

a. pH is inversely related to free $[H^+]$: Low pH has a high $[H^+]$, high pH has a low $[H^+]$
b. pH scale goes from 0 to 14.

2. The Henderson-Hasselbalch equation relates pH to the acid dissociation constant and is expressed as:

$$pH = pK_a + \log((A^-)/(HA))$$

When $[A^-] = [HA]$, the solution $\mathbf{pH = pK_a}$.

3. The **dissociation constant (K)** for water is equal to:

$$K = (H^+) \times (OH^-)/(H_2O)$$

a. The ion product of water (Kw) is $Kw = [H^+] \times [OH^-]$.
b. In any aqueous solution, the product of $[H^+]$ and $[OH^-]$ is a constant.

4. The dissociation constant determines the availability of free $[H^+]$ for different acids or bases with the same molarity.

5. The pH of arterial blood averages 7.4, which is slightly higher than the pH of venous blood (7.35). Most body fluids fall within a pH range of 5-8 units.

E. Buffers promote stability of pH.

1. A **pH buffer** minimizes a change in pH when either acid or base is added to a solution.

2. A chemical buffer consists of a weak acid and its conjugate base or a weak base and its conjugate acid.

3. The equilibrium expression for dissociation of an acid can be written in the **Henderson-Hasselbalch** equation form:

$$pH = pK_a + \log((\text{conjugate base})/(\text{acid}))$$

a. For a specific buffer, the pH change for a given amount of added acid or base will be least near the pKa of the buffer.
b. The slope of a **titration curve** is an index of the effectiveness of the buffer in resisting a change in pH and is flattest when $pH = pK_a$ of the buffer.

4. The effectiveness of a buffer in minimizing pH changes depends on two factors:
a. The pK_a of the buffer in relation to the desired pH.
b. The amount or concentration of buffer present.

F. Hydrogen ions are produced in metabolic reactions.

1. The metabolism of food in our diet is usually the major source of hydrogen ions.

2. Carbon dioxide reacts with water to produce carbonic acid (H_2CO_3), which dissociates with H^+ and HCO_3^-.
a. As long as CO_2 is expired, no acid burden is imposed on the body.
b. If carbohydrates and fatty acids are completely oxidized, no acid burden is imposed on the body, provided that CO_2 does not accumulate.
c. Incomplete oxidation increases the acid burden on the body.

3. The metabolism of dietary proteins results in production of strong acids.
a. To some extent these acid-producing reactions are neutralized by acid-consuming reactions.
b. An excess of 50 – 100 mEq H^+ per day is produced on the usual American diet.

II. Regulation of Extracellular pH: Buffering Mechanisms in the Body

A. Many chemical buffers stabilize extracellular fluid pH.

1. pH is stabilized in the body by:
a. Chemical buffers in extracellular and intracellular fluids and in bone.
b. The lungs, which dispose of CO_2.
c. The kidneys, which buffer blood pH by excreting hydrogen ions.

2. **Chemical pH buffers** include phosphates, proteins, and the bicarbonate $-CO_2$ system.
a. Chemical buffers consist of conjugate acid-base pairs that bind or release H^+ when acid or base is added.
b. Phosphate with pK of 6.8 is a good buffer for stabilizing blood pH close to its normal value of 7.4.
c. Inorganic phosphate has a limited capacity to buffer because of its low

concentration in plasma and ECF.
d. Phosphate salts in bone and in organic phosphate compounds in the cell are effective buffers.
e. Inorganic phosphate is an effective buffer in the urine.

3. Proteins constitute the largest amount of buffer because they contain many different ionizable groups.

4. The **bicarbonate** $-CO_2$ system is of special importance in pH buffering in extracellular fluids and is represented by the following reaction:

$$CO_2 + H_2O \Leftrightarrow H_2CO_3 \Leftrightarrow H^+ + HCO_3^-$$

a. The concentration of dissolved CO_2 in the plasma is related to PCO_2.
b. The concentration of H_2CO_3 is directly proportional to the P_{CO2} by way of dissolved CO_2.
c. The pK of bicarbonate $-CO_2$ buffer pair is 6.10, but the buffer pair operates in an open system. CO_2 can be exhaled, which lowers its level in plasma.
d. The bicarbonate $-CO_2$ system is effective because of the ability of the body to change the amounts of components of this particular buffer pair.

5. In any fluid compartment, all the buffer systems are in equilibrium with the same H^+ concentration, a concept called the isohydric principle.

B. *The lungs and kidneys normally stabilize extracellular fluid pH.*

1. CO_2 is usually expired at the same rate as it is produced from metabolism, and the PCO_2 remains near normal conditions.
a. In a state of **hyperventilation**, CO_2 is exhaled at a rate greater than the rate at which it is produced in the body.
b. This results in the bicarbonate $-CO_2$ equation shifting to the left, and the blood becoming more alkaline.
c. During **hypoventilation**, CO_2 is exhaled at a rate slower than the rate at which it is produced.
d. This results in the bicarbonate $-CO_2$ equation shifting to the right and blood becoming more acidic.

2. The concentration of a fixed acid is not affected by lung function. Addition of fixed acid to the blood results in increased pulmonary ventilation.

3. **Peripheral and central chemoreceptors** detect changes in blood pH, which stimulates the appropriate changes in ventilation.

4. Increased ventilation lowers arterial blood P_{CO2} and carbonic acid concentration thus minimizing a fall in blood pH.

5. Buffering of acid-base disturbances by the respiratory system is rapid but coarse; renal responses to their types of disturbances are slower but more effective to return blood pH to normal.

6. The kidneys, acting as a buffering system, get rid of acid or base in the urine when there is an excess or deficit of hydrogen ions in the body. Most of the H^+ excreted in the urine are combined with urinary buffers, such as ammonia and inorganic phosphate.

7. Three processes are involved in **urinary acidification**:
a. Reabsorption of filtered bicarbonate.
b. Excretion of titratable acid.
c. Excretion of ammonia.
d. All three involve H^+ secretion and are associated with addition of bicarbonate to the peritubular capillary blood by kidney tubules.

8. Urine is acidified by the secretion of H^+ by the tubular epithelium in exchange for Na^+.
a. H^+ combines with $HCO3^-$ to form H_2CO_3, which undergoes dehydration to CO_2 and H_2O by carbonic anhydrase in the membrane.
b. In the cell, this reaction is catalyzed by cellular carbonic anhydrase.
c. H_2CO_3 dissociates to H^+, which are secreted into the lumen, and $HCO3^-$, which exits through the basolateral membrane to plasma.
d. The filtered bicarbonate is reabsorbed indirectly by H^+ secretion.

9. Filtered phosphate is incompletely reabsorbed, and the unreabsorbed phosphate acts as a pH buffer in the urine.

10. Ammonia produced by kidney tubule cells diffuses or is actively secreted into the urine, where it combines with H^+ to form the ammonium ion.
    a. The ammonium ion is trapped in the urine.
    b. For each H^+ exerted as part of the ammonium ion, an equivalent quantity of new bicarbonate is added to the blood.
    c. A high blood P_{CO2} favors generation and increased excretion of H+ combined with titratable acids and ammonia and more complete reabsorption of filtered bicarbonate.
    d. A low blood P_{CO2} results in a lower rate of hydrogen ion secretion, less complete reabsorption of bicarbonate, and the excretion of an alkaline (bicarbonate-rich) urine.
    e. The kidneys regulate plasma pH by changing plasma bicarbonate concentration in the appropriate direction.

III. Disturbances of Acid-Base Balance

~ The four fundamental types of acid-base disturbance are respiratory acidosis, respiratory alkalosis, metabolic acidosis, and metabolic alkalosis.

~ These four disturbances of acid-base balance are each accompanied by characteristic changes in arterial blood pH, PCO_2 and $[HCO_3^-]$.

A. Accumulation of CO_2 results in respiratory acidosis.

1. **Respiratory acidosis**, characterized by accumulation of CO_2 owing to inadequate alveolar ventilation, may result from:
    a. Insufficient neural drive for ventilation.
    b. Inadequate movements of respiratory muscles or thoracic cage.
    c. Airway obstruction or pulmonary disease.

2. Plasma concentrations of HCO_3^- and H^+ increase.

3. The kidneys compensate by secreting hydrogen ions into tubular urine and reabsorbing all the filtered bicarbonate. Hydrogen ions combined with urinary buffers are excreted.

4. The bicarbonate added to the blood diminishes the severity of the acidemia.

*B. Excessive loss of CO_2 results in **respiratory alkalosis**.*

1. The excessive loss of CO_2 by alveolar hyperventilation leads to a fall in plasma hydrogen ion concentration, an increase in blood pH, and a decline in plasma bicarbonate.

2. Hyperventilation may be caused by voluntary effort, anxiety, or stimulation of respiratory centers by an abnormal influence acting directly on the brain.

3. Chemical buffers release H^+ with most chemical buffering occurring in the cells.

4. The kidneys compensate by excreting bicarbonate in the urine and a decreased hydrogen in secretion by the tubular epithelium.

C. Metabolic (nonrespiratory) acidosis results from gain of fixed acid or loss of bicarbonate.

1. **Metabolic acidosis** can result from:
    a. Renal failure.
    b. Uncontrolled diabetes mellitus.
    c. Inadequate circulation.
    d. Heavy exercise.
    e. Ingestion of certain acidifying agents.
    f. Loss of HCO_3^- due to diarrhea or abnormal excretion.

2. Chemical buffering occurs in extracellular fluids, cells, and bone.

3. Respiratory compensation is rapid but usually does not restore blood to normal pH.

4. Renal compensation is slower but more complete.
    a. The kidneys increase the urinary output of H^+ combined with urinary buffers (phosphate, ammonia).
    b. Filtered HCO_3^- is reabsorbed.

D. Metabolic (nonrespiratory) alkalosis results from gain of strong base or bicarbonate or loss of acid.

1. **Metabolic alkalosis** can result from:
    a. Ingestion of excess sodium bicarbonate.
    b. Oxidation of organic anions.
    c. Vomiting.

2. Metabolic alkalosis leads to an increase in

plasma bicarbonate, a rise in pH, and a compensatory rise in PCO_2.

3. Chemical buffers release H^+ to limit the alkaline shift in pH.

4. Respiratory compensation is alveolar hypoventilation, but this is limited.

5. Renal compensation involves increased excretion of bicarbonate and a fall in plasma concentration of HCO_3^-.

E. *Clinical evaluation of acid-base status involves a history, physical examination, and arterial blood chemistry values.*

1. Necessary laboratory data include blood pH and PCO_2 and plasma bicarbonate concentration.

2. To evaluate blood chemistry values, normal values for blood variables must be known.

3. A person may develop two independent disturbances of acid-base balance, i.e., a mixed acid-base disturbance.

IV. Intracellular pH Regulation

A. *Intracellular pH is closely regulated.*

1. Intracellular pH affects many physiological activities, such as muscle contraction, opening and closing of membrane ion channels, secretion, and cell division.

2. Intracellular $[H^+]$ is kept low by various transport mechanisms that remove H^+ or add HCO_3^-, such as cellular membrane Na^+/H^+ exchanger and bicarbonate transporting systems (e.g., Na-independent and Na-dependent Cl^-/HCO_3^- exchangers).

3. Cellular proteins and organic phosphates serve as buffers.

4. H^+ may be sequestered by various organelles, e.g., through the activity of H^+-ATPase in endosomes.

Key Terms

acid	dissociation
amphoteric substance	Henderson-Hasselbalch
association	metabolic acidosis
base	metabolic alkalosis
bicarbonate $-CO_2$ system	respiratory acidosis
buffer	respiratory alkalosis
chemoreceptor	

Testing Your Knowledge

I. Matching. Select the correct term from the Key Terms.

________ 1. donates hydrogen ions

________ 2. accepts or binds hydrogen ions

________ 3. describes substance that can donate and accept hydrogen ions

________ 4. describes reaction $HA \rightarrow H^+ + A^-$

________ 5. describes reaction $H^+ + A^- \rightarrow HA$

________ 6. substance that minimizes a change in pH when acid or base is added to solution

________ 7. detects change in blood pH

________ 8. characterized by accumulation of CO_2

________ 9. characterized by loss of CO_2

________ 10. characterized by gain of fixed acid or loss of bicarbonate

II. Dichotomous. Circle the correct choice from the paired terms in parentheses.

1. A(n) (acid/base) is a proton donor, while a(n) (acid/base) is a proton acceptor.

2. At a pH of 3, the (ammonium/carboxylate) group of arginine will (donate/accept) a hydrogen ion.

3. Metabolism of (proteins/carbohydrates) produces strong acids and increased $[H^+]$.

4. The reaction $H_2CO_3 \rightarrow CO_2 + H_2O$ is a (hydration/dehydration) reaction.

5. In a closed system, P_{CO2} is (greater/less) than P_{CO2} in an open system, and pH would be (raised/lowered) compared to an open system.

6. Alveolar ventilation and P_{CO2} are (directly/ inversely) related so that when a person hyperventilates, arterial blood pH becomes more (acidic/basic).

7. A low blood P_{CO2} results in (lower/higher) rate excretion of H^+ and (lower/higher) reabsorption of filtered HCO_3^-.

8. (NH_3/NH_4) diffuses readily through cell membrane.

9. Under conditions of excess acid in the body, the reaction glutamine → NH_4^+ + ketoglutarate^{2-} is (increased/decreased) in tubular epithelial cells.

10. In a person with respiratory alkalosis, the log term in the Henderson-Hasselbalch equation is (greater/less) than this term would be in a person with respiratory acidosis.

III. True/False. In the space to the left of the number write T for true or F for false.

_______ 1. 10 M HCl has a low dissociation constant.

_______ 2. A 1-M solution of "X" at pH 6 has 10 times more H^+ than a 1-M solution of "X" at pH 5.

_______ 3. Proteins such as hemoglobin are poor chemical buffers.

_______ 4. The reaction $HCO_3^- + H^+ \rightarrow H_2CO_3$ is favored in erythrocytes as $[HCO_3^-]$ increases.

_______ 5. The slope of a titration curve of a phosphate buffer is greatest around its pKa.

_______ 6. Hydrogen ions are excreted in combination with NH_3 or HPO_4^{2-}.

_______ 7. Most of the hydrogen ions are secreted by distal convoluted tubules through the action of a Na^+/H^+ exchanger.

_______ 8. An absence of carbonic anhydrase in the brush border membrane would have a greater effect on urinary acidification by reabsorption of filtered HCO_3^- than by excretion of ammonia.

_______ 9. Under a condition of excess base in the body, the kidneys increase the output of H^+ to lower the pH to normal values.

_______ 10. Respiratory compensation is more effective than renal compensation under conditions of either metabolic acidosis or alkalosis.

IV. Sentence Completion. Write your response in the space provided.

1. For 0.1 M acetic acid, the concentration of free H^+ is ____________.

2. The pKa of 0.1 M lactic acid is 3.9. Half of the acid will be present in its dissociated form at pH ____________.

3. You are studying the activity of pepsin. To simulate its environment you need to use a buffer with a pKa around ____________.

4. The reaction $CO_2 + H_2O \Leftrightarrow H_2CO_3$ is catalyzed by the enzyme ____________.

5. In the equation $pH = 6.8 + \log ([HPO_4^{2-}]/[H_2PO_4^-])$, the dihydrogen phosphate is a(n) ____________ and the monohydrogen phosphate is a(n) ____________.

6. If a person ventilates at a rate of 30 per minute, the equation $CO_2 + H_2O \Leftrightarrow H_2CO_3 \Leftrightarrow H^+ + HCO_3^-$ will shift to the __________.

7. A decrease in blood pH caused by infusion of fixed acid is detected by (peripheral/ central) chemoreceptors, but an increase in $[H^+]$ in blood caused by hyperventilation is detected by (peripheral/central) chemoreceptors.

8. The new HCO_3^- produced by the urinary acidification process of formation of NH^4 is derived from ____________ in the cells of the ____________.

9. If 2 mEq of H^+ are excreted as titratable acid,

new HCO_3^- are added to blood.

10. To maintain intracellular pH below equilibrium, ___________ ions are actively transported from the cell by the ___________ exchanger.

V. Multiple Choice. Circle the letter of the best response.

1. Hydrogen ions are produced under all of the following conditions except:
   a. strenuous exercise
   b. starvation
   c. reduced respiration rate
   d. high protein diet
   e. a vegetarian diet

2. The kidneys play an important role in regulation of body fluid pH by excreting all of the following except:
   a. sulfate anions
   b. hydrogen ions
   c. phosphate anions
   d. bicarbonate anions
   e. chloride anions

3. Consider the Henderson-Hasselbalch equation $pH = 6.10 + \log([HCO_3^-]/0.03\ P_{CO2})$. If the CO_2 dissolved in pulmonary capillaries is 1.2 mmol/liter and $[HCO_3^-] = 24$ mmol per liter, the pH in an open system is:
   a. 6.2
   b. 6.7
   c. 7.0
   d. 7.2
   e. 7.4

4. Consider the equation $pH = 6.8 + \log([HPO_4^{2-}]/[H_2PO_4^-])$. If the ratio of monohydrogen phosphate to dihydrogen phosphate is 0.5, the pH will be:
   a. 6.5
   b. 6.8
   c. 7.1
   d. 7.4
   e. none of the above

5. If a fixed acid is added to the blood:
   a. alveolar ventilation is decreased
   b. alveolar PCO_2 increases
   c. $[H_2CO_3]$ is decreased
   d. CO_2 dissolved in pulmonary capillaries increases
   e. plasma pH increases

6. All three processes in urine acidification involve all the following except:
   a. reabsorption of filtered HCO_3^-
   b. cotransport of Na^+ and HCO_3^- through the basolateral membrane of tubular epithelial cells
   c. secretion of H^+
   d. formation of H_2CO_3 in tubular epithelial cells
   e. none of the above

7. The source of H^+ in urinary acidification by the excretion of ammonia is:
   a. HCO_3^-
   b. H_2CO_3
   c. glutamine
   d. H_2O
   e. $H_2PO_4^-$

8. Respiratory acidosis is characterized by:
   a. a shift to the left of the reaction $CO_2 + H_2O \Leftrightarrow H_2CO_3 \Leftrightarrow H^+ + HCO_3^-$
   b. an increase in blood pH
   c. an increase in P_{CO2}
   d. increased excretion of HCO_3^- by the kidneys
   e. alveolar hyperventilation

9. A common feature of both metabolic acidosis and metabolic alkalosis is:
   a. chemical buffering in extracellular fluids
   b. respiratory compensation by alveolar hyperventilation
   c. renal compensation by increased secretion of H^+
   d. decreased plasma bicarbonate concentration
   e. decreased P_{CO2} in arterial blood

10. Consider the following systemic arterial acid-base data for a person who has had diarrhea for two days: pH 7.2, P_{CO2} 40 mm Hg, and plasma $[HCO_3^-]$ 22 mEq per liter. This person most likely has an acid-base disturbance of:
   a. respiratory acidosis
   b. respiratory alkalosis
   c. metabolic acidosis
   d. metabolic alkalosis
   e. a combination of respiratory and metabolic alkalosis

VI. Complete the Table.

For each condition, indicate the effect (increase, decrease, none) on secretion of H^+, the reabsorption of bicarbonate, and blood pH.

Condition	Water Intake	ADH Release
1. Ingestion of excess salt		
2. Decreased function of adrenal cortex		
3. Excessive sweating		
4. Activation of thirst sensation		
5. Decrease in plasma osmolality		
6. Cellular dehydration		
7. Hypovolemia		
8. Presence of renin in the plasma		
9. Presence of ADH		
10. Decrease in effective arterial blood volume		

Chapter 26

Calcium, Phosphate, and Bone Metabolism

Chapter Learning Objectives

1. Describe the distribution of calcium and phosphate in tissue and describe physiological actions of phosphate and calcium.
2. Distinguish between filterable and bound calcium.
3. Describe the homeostatic mechanisms involved in the metabolism of calcium and phosphate.
4. Describe where and how calcium and phosphorus are absorbed in the gastrointestinal tract.
5. Describe where and how calcium and phosphorus are reabsorbed in the kidney.
6. Outline the processes of bone synthesis and resorption.
7. Give the functions of osteoblasts, osteoclasts, osteocytes, and chondrocytes.
8. Account for the hormonal regulation of calcium and phosphate. (a) Describe nonhormonal mechanisms for calcium and phosphate regulation. (b) Describe the actions of parathyroid hormone, calcitonin, and the active forms of vitamin D.
9. Indicate how vitamin D affects the body.
10. Describe various abnormalities of the bone and the factors contributing to their development.

Expanded Chapter Outline

I. Overview: Functions of Calcium and Phosphate

A. Calcium and phosphorus are distributed differently in tissues.

1. Calcium plays a role in a number of different physiological processes:
   a. Maintaining normal sodium permeability in nerves.
   b. Triggering the endings at the neuromuscular junction.
   c. Acting as a signal for excitation-contraction coupling in muscle.
   d. Acting as an intracellular signal or second messenger for normal activity.
   e. Serving as a cofactor for enzymes and hormones.
   f. Playing a structural role as a component of bone.

2. Phosphorus is involved in many physiological actions, including:
   a. Functioning as part of the intracellular buffer system.
   b. Serving as a constituent of nucleic acids, phospholipids, and other macromolecules.
   c. Regulating metabolic processing by phosphorylation and dephosphorylation of key regulatory enzymes.

3. About 99% of total body calcium is located in bone. The rest is in muscle (0.3%) and in cells and extracellular fluid.

4. About 85% of total body phosphate is located in bone. Also contained in bone are other inorganic minerals, water, and carbonate, which can be released to combat acidosis.

5. Bone provides a third line of defense in acid-base regulation along with the lungs and kidneys.

B. Calcium and phosphorus exist in several forms in the blood.

1. The two major forms of calcium in the plasma are **filterable calcium**, which can be subdivided into ionized and complexed forms, and **protein-bound calcium**.

2. Filterable calcium (about 60% of total calcium in the plasma, or 6 mg/100 ml) comprises those forms readily filtered in the glomerulus and reabsorbed in the renal tubules.

3. **Ionized calcium** ($Ca2+$), the physiologically active form, is about 50% of total calcium in the plasma, can diffuse across plasma mem-

branes, and is the form that is regulated in the plasma.

4. **Complexed calcium**, about 10% of total calcium in plasma, may be bound to bicarbonate or citrate.

5. **Protein-bound calcium**, about 40% of total calcium in the plasma, is not filtered in the kidney and is bound mainly to albumin.

6. The plasma phosphate concentration fluctuates daily between 3.0 and 4.5 mg/100 ml.

7. Phosphorus exists as **orthophosphate** (PO_4) in the plasma, about 80% as HPO_4^{2-} and 20% as $H_2PO_4^-$.

8. In its ionized form, phosphorus diffuses readily across membranes and is filterable in the kidney.

9. Phosphorus may also be present in organic form.

II. Pathways of Calcium and Phosphate Homeostasis

A. Understanding pathways of calcium homeostasis provides the basics for further understanding of how plasma calcium is regulated.

1. The diet is the ultimate source of calcium and phosphorus.
   a. Plasma concentrations are determined by the intestines, bone, and kidneys.
   b. The majority of ingested calcium is not absorbed, and some body calcium is lost to the intestines.
   c. Ninety-nine percent of filtered calcium is reabsorbed and returned to the blood.

2. There is a daily dynamic flux of calcium into and out of bone.

3. The processes of calcium absorption from the intestines, calcium reabsorption by the kidneys, and bone mineral formation and resorption are precisely controlled.

B. Understanding pathways of phosphate homeostasis also provides important basic concepts.

1. Most of ingested phosphate is absorbed in the intestines, but a small amount is also secreted into the intestines each day.

2. Phosphate ion is reabsorbed from the fiitrate and returned to the plasma.

3. There is a daily dynamic flux of phosphate into and out of the bone.

III. Mechanisms Involved in Calcium and Phosphate Homeostasis

A. Calcium and phosphate absorption in the gastrointestinal tract differ in magnitude.

1. Calcium is absorbed by active transport, which occurs predominantly in the duodenum and jejunum, and by simple diffusion, which occurs predominantly in the ileum.
   a. Active transport of calcium exhibits saturability and requires energy.
   b. Simple diffusion is directly related to the concentration gradient of calcium between the lumen and the plasma.
   c. The relative importance of active transport and passive diffusion depends on calcium intake.
   d. Calcium uptake from the intestines is regulated through the regulation of active calcium transport mechanisms.

2. Phosphate absorption occurs by active transport and passive diffusion predominantly in the small intestine.
   a. Active transport of phosphate is in part coupled with that of calcium.
   b. Phosphate absorption is only slightly regulated, if at all.

B. Calcium and phosphate reabsorption in the kidney differ by tubular location and extent.

1. Reabsorption of calcium occurs by active transport in the proximal and distal tubules and by passive diffusion from the ascending limb of the loop of Henle.

2. Calcium reabsorption is regulated in the distal tubule by the action of parathyroid hormone.

3. About 85% of filtered phosphate is reabsorbed by active transport processes operating primarily in the proximal tubule but also in the distal tubule.

4. Phosphate reabsorption in the proximal tubule is inhibited by parathyroid hormone.

C. Calcium and phosphate homeostasis in the bone is a highly regulated process.

1. Mature bone consists of inorganic bone mineral deposited in a framework of organic support material.
   a. The mineral is largely calcium phosphate in the form of **hydroxyapatite** crystals.
   b. The **organic matrix**, known as **osteoid**, is composed of collagen (95%) and ground substance (5%), which consists of a mixture of various proteoglycans.

2. The three types of bone cells involved in bone formation and reabsorption are as follows.
   a. **Osteoblasts** synthesize osteoid, which is deposited on the surface of the bone.
   b. **Osteocytes** are osteoblasts that have lost their bone-forming capability and communicate by canaliculi.
   c. **Osteoclasts** reabsorb bone by secreting both acids and proteolytic enzymes into extracellular space adjacent to the bone.

3. The fetal skeleton is largely composed of **cartilage**, which serves as the template for the development of the bony skeleton.

4. The process of cartilage replacement begins in the middle of the bone and progresses outward toward the ends, where the epiphyseal plate is located.

5. At the **epiphyseal plate**, the zone of active bone formation moves forward as a result of:
   a. The continuing process of cartilage synthesis by **chondrocytes**.
   b. **Calcification** and later erosion of cartilage.
   c. Osteoblast invasion and replacement of cartilage by bone.

6. Around puberty the epiphyseal plates of long bones become unresponsive to hormonal stimuli.

7. I6F-1, **somatomedin**, is produced in the liver in response to growth hormone.
   a. It stimulates chondrocyte activity.
   b. Insulin and thyroid hormones also stimulate chondrocyte activity and promote bone growth.

8. Remodeling of bones is a continuous process and serves to meet the changing mechanical needs occurring throughout life.

IV. Regulation of Calcium and Phosphate in Plasma

A. Nonhormonal mechanisms for calcium and phosphate regulation provide a rapid but low-capacity system of regulation.

1. About 1% of calcium in bone exists in a readily exchangeable form that can move rapidly between bone and extracellular fluid.

2. The free ionized calcium in extracellular fluid exists in chemical equilibrium with calcium present in the crystal matrix.

3. A decrease in free calcium in extracellular fluid directly results in calcium movement from bone to extracellular fluid.

B. Hormonal mechanisms for calcium and phosphate regulation provide for long-term regulation.

1. Three hormones are involved in the regulation of calcium ion concentrations in the extracellular fluid:
   a. **Parathyroid hormone.**
   b. **Calcitonin.**
   c. **Metabolites of vitamin D.**

2. The actions of the three hormones are intimately interrelated by a series of feedback loops involving ionized calcium, Ca^{2+}, as the regulated factor.

3. **Parathyroid hormone (PTH)** produced by the parathyroid glands is released in response to a decrease in the concentration of free ionized calcium in the plasma.

4. **Calcitonin (CT)** produced by parafollicular cells in the thyroid is secreted in response to an increase in the plasma concentration of ionized calcium. It is also secreted in response to several gastrointestinal hormones, particularly gastrin.

5. **Metabolites of vitamin D**, such as 1, 25-dihydroxyvitamin D_3, are the hormonally active forms that regulate plasma calcium concentrations.
   a. These metabolites are formed from D vitamins, D_3 and D_2, which are enzymatically modified and activated in the liver and kidney.

b. The primary site of regulation of the formation of the active form is the proximal tubule, where the enzyme 1 α-hydroxylase catalyzes the conversion of 25-hydroxyvitamin D_3 to 1, 25-dihydroxyvitamin D_3, the active form.
c. The activity of the enzyme is increased by PTH and indirectly as a result of changes in plasma calcium concentration as well as by a decrease in plasma phosphate concentration.

6. PTH is essential in the regulation of plasma calcium concentration.

7. The effects of PTH on the kidney include:
   a. Increasing the activity of 1 -hydroxylase to form the active metabolite of vitamin D.
   b. Stimulation of the active transport mechanism for calcium reabsorption in the distal tubules.
   c. Decreasing the proximal tubular reabsorption of phosphate.

8. The effects of PTH on the bone include:
   a. Stimulation of dissolution of bone by increasing activity of osteoclasts.
   b. Stimulation of maturation of immature osteoclasts.
   c. Decreasing collagen synthesis in osteoblasts.

9. **Calcitonin** provides a "fine-tuning" for the regulation of calcium and phosphate metabolism

10. The effects of calcitonin include:
    a. Decreasing tubular reabsorption of calcium and phosphate.
    b. Decreasing resorption of calcium by inhibiting activity of osteoclasts.

11. The effects of the **active form of vitamin D** include:
    a. Increasing the tubular reabsorption of both calcium and phosphate.
    b. Promoting the actions of PTH in the bone.
    c. Stimulation of both calcium and phosphate absorption in the small intestine.

V. Abnormalities of Bone Mineral Metabolism

A. *Osteoporosis results from an equivalent loss of bone mineral and organic matrix.*

1. The two major categories of metabolic bone disease are osteoporosis and osteomalacia/rickets.

2. **Osteoporosis** is a disease in which there is a reduction in overall bone mass with equal losses of both mineral and organic matrix.

3. Among the factors contributing to osteoporosis are:
   a. Long-term calcium deficiency.
   b. Pronounced vitamin C deficiency.
   c. Immobilization.
   d. Advancing age.

B. *Osteomalacia and rickets result from inadequate mineralization of new bone.*

1. Both diseases are characterized by inadequate mineralization of new bone matrix, such that the ratio of mineral to organic matrix is lower than normal.

2. The disorder is called rickets in children and osteomalacia in adults.

3. Rickets and osteomalacia are most commonly caused by a deficiency in vitamin D activity as a result of one of three general causes:
   a. Combined dietary deficiency of vitamin D and insufficient exposure to sunlight.
   b. Impairments resulting from liver disease, kidney disease, or absence of PTH in the conversion of vitamin D in the liver and kidney to its hormonally active form.
   c. Inability of the active form of vitamin D to act in its target tissues.

Key Terms

calcification
calcitonin
chondrocyte
collagen
complexed calcium
1,25-dihydroxyvitamin D3
epiphyseal plate
filtered calcium
hydroxyapatite crystal
1 α-hydroxylase
hypocalcemia
ionized calcium
orthophosphate
osteoblast
osteoclast
osteocyte
osteoid
osteomalacia
osteoporosis
parathyroid hormone
phosphaturia
protein-bound calcium
proteoglycan
rickets
tetany

Testing Your Knowledge

I. Matching. Select the correct term from the Key Terms.

_________ 1. represents 60% of total body calcium

_________ 2. results from apostic contractions of muscles

_________ 3. mineral fraction of bone

_________ 4. organic matrix of bone

_________ 5. synthesize and secrete organic matrix of bone

_________ 6. involved in bone resorption

_________ 7. site of elongation or growth of bone after birth

_________ 8. synthesize cartilage

_________ 9. synthesized and secreted in response to decrease in plasma calcium; most important regulator for calcium

_________ 10. synthesize by parafollicular cells

_________ 11. active form of vitamin D

_________ 12. component of organic matrix of bone

_________ 13. results from increased phosphate excretion

_________ 14. characterized by decrease in over all bone mass

_________ 15. characterized by inadequate mineralization

II. Dichotomous. Circle the correct choice from the paired terms in parentheses.

1. (Phosphorus/Calcium) plays an important role as an intracellular buffer.

2. About 4.8 mg/100 ml plasma calcium is in the (ionized/complexed) form.

3. When dietary calcium intake is low, (active transport/passive diffusion) predominates in the intestine.

4. For both calcium and phosphorus, most reabsorption occurs in the (proximal/distal) tubule by (active transport/passive diffusion).

5. The effect of parathyroid hormone is to (increase/decrease) plasma concentration of calcium and to (increase/decrease) phosphorus plasma concentration.

6. In a 5-year-old child, (osteoblasts/osteoclasts) would be found in the epiphyseal plates.

7. Calcitonin is released in response to a(n) (increase/ decrease) in the concentration of plasma calcium, while parathyroid hormone is secreted in response to a(n) (increase/decrease) in plasma calcium concentration.

8. Calcitonin decreases bone (formation/resorption) by inhibiting the activity of (osteoclasts/osteoblasts).

9. In the presence of 1,25-$(OH)_2$ vitamin D_3, the effect of parathyroid hormone is (greater/less) than it would be in its absence.

10. In osteoporosis, bone resorption is (greater/less) than bone formation.

III. True/False. In the space to the left of the number write T for true or F for false.

_______ 1. A muscle cell contains more grams of phosphorus than calcium.

_______ 2. Calcium bound to albumin makes up 50% of filterable calcium.

_______ 3. In the small intestine, active transport of phosphorus is coupled to calcium transport.

_______ 4. In the femurs of a fetus, cartilage turns into bone.

_______ 5. Hormone mechanisms for calcium regulation permit more rapid mobilization of calcium than nonhormonal mechanisms.

_______ 6. Plasma calcium concentration and secretion of parathyroid hormone are inversely related.

_______ 7. A decrease in phosphate causes an increase in the activity of 1 α-hydroxylase, which results in a decrease in the production of 1,25-$(OH)_2$-vitamin D_3.

_______ 8. Parathyroid hormone acts directly on the gastrointestinal tract to increase absorption of calcium and phosphate.

_______ 9. A person with a deficiency of receptors for 1,25-$(OH)_2$-vitamin D_3 would have normal bone formation if sufficient dietary calcium were ingested.

_______ 10. The receptor for 1,25-$(OH)_2$-vitamin D_3 is located in the cell's membrane.

IV. Sentence Completion. Write your response in the space provided.

1. More than 990 g of ___________ and 4.25 g of ___________ are found in a person's bones.

2. At normal blood pH, about 80% of phosphate exists as (the) ___________ and 20% as the ___________.

3. About 1% of ___________ and 15% of ___________ are excreted.

4. Ninety-five percent of the organic matrix of the bone is composed of ___________ and 5% is composed of ___________.

5. The final enzymatic modification of vitamin D_3 to form the hormonally active form occurs in the ___________.

6. ___________ may be obtained from the diet or formed in the skin by the action of ultraviolet light.

7. In bone, parathyroid hormone stimulates the activity of ___________.

8. A chicken bone soaked in 5% acetic acid for five days becomes bendable. This change results from loss of the ___________ constituent of bone.

9. In a person with osteomalacia, the ratio of mineral to organic matrix is ___________ than in a healthy person.

10. IGF-1 stimulates ___________ to promote bone growth in the ___________ region.

V. Multiple Choice. Circle the letter of the best response.

1. All the following activities depend on calcium except:
   a. release of acetylcholine
   b. blood clotting
   c. intracellular signaling
   d. synthesis of ATP
   e. formation of bone

2. A person with chronic acidosis would experience a loss of ___________ from his/her bones.
   a. calcium
   b. magnesium
   c. carbonate
   d. phosphorus
   e. sodium

3. The form of calcium that is strictly regulated in the plasma is:
   a. Ca^{2+}
   b. Ca
   c. Ca-albumin
   d. $CaCO^3$
   e. $Ca^3(PO_4)^2$

4. For each of the physiological activities listed, the amount of calcium is greater than the amount of phosphorus except:
   a. dietary input
   b. flow into and out of bone
   c. concentration in extracellular fluid
   d. loss in feces
   e. intracellular concentration

5. Bone cells active in reabsorption are called:
   a. osteoblasts
   b. osteocytes
   c. osteoclasts
   d. osteoids
   e. chondrocytes

6. All of the following physiological activities are common effects of both parathyroid hormones and 1, 25-$(OH)_2$-vitamin D_3 except:
   a. increase plasma calcium concentration
   b. increase reabsorption of calcium
   c. increase reabsorption of phosphorus
   d. increase resorption of bone
   e. none of the above

7. A deficiency of 1 -hydroxylase has its greatest effect on the formation of:
   a. cholecalciferol
   b. ergocalciferol
   c. 25-OH-vitamin D_3
   d. 1, 25-dihydroxyvitamin D_3
   e. calcitonin

8. Which of the following (hormone; target tissue; effect) is mismatched?
   a. parathyroid hormone; kidney; increased activity of 1 α-hydroxylase
   b. calcitonin; bone; increased resorption of calcium
   c. 1, 25-$(OH)_2$-D_3; kidney; increased reabsorption of phosphate
   d. 1, 25-$(OH)_2$-D_3; gastrointestinal tract; increased absorption of calcium
   e. parathyroid hormone; bone; increased resorption of bone

9. Plasma concentration of calcium may be increased by all of the following except:
   a. increased bone resorption by 1,25-$(OH)_2$-vitamin D_3
   b. increased absorption of calcium from the gastrointestinal tract by calcitonin
   c. increased bone resorption by parathyroid hormone
   d. increased reabsorption in the kidney by parathyroid hormone
   e. increased reabsorption in the kidney by 1, 25-$(OH)_2$-vitamin D_3

10. 1, 25-$(OH)_2$-vitamin D_3 exerts all of the following effects except:
    a. decreased production of parathyroid hormone
    b. decreased production of interleukin-2
    c. promotes macrophage differentiation
    d. antiproliferative action
    e. none of the above

VI. Complete the Table.

For each condition, indicate the effect (increase, decrease, none) on calcium or phosphate reabsorption and on plasma calcium level.

Condition	Plasma Calcium Level	Effect on Calcium Reabsorption	Effect on Phosphate Reabsorption
1. Presence of PTH			
2. Presence of calcitonin			
3. Presence of 1, 25-$(OH)_2$ vitamin D_3			
4. Absence of PTH			
5. Living in a cave			

Chapter 27

Regulation of Body Temperature

Chapter Learning Objectives

1. Distinguish between homeotherms and poikilotherms.
2. Define metabolic rate and describe several factors affecting metabolic rate.
3. Describe methods for measuring metabolic rate.
4. Indicate how metabolic rate influences body temperature.
5. Explain the balance between heat production and heat loss.
6. Describe four physical processes by which the body loses heat and describe heat transfer within the body.
7. Explain the differences in rectal, oral, and axillary measurements. Describe various methods for measuring core body and skin temperatures.
8. List several factors affecting body temperature, and describe how each acts.
9. Point out examples of circadian rhythms in body temperature.
10. Discuss the body's response to abnormal temperatures. Describe the cause of and effect on the body of fever, heat stroke, heat exhaustion, and hypothermia.

Expanded Chapter Outline

I. The Balance Between Heat Production and Heat Loss

A. Regulation of body temperature involves a balance between heat production and heat loss.

1. Thermal homeostasis is maintained by physical and physiological regulatory mechanisms that operate to keep heat production and heat loss approximately equal.

2. Heat is produced continuously as a by-product of the catabolism of food.

> Food energy = chemical storage + external work + metabolic heat

   a. Eighty percent of food energy is liberated as metabolic heat.
   b. Twenty percent is available to do external work.

3. **Metabolic rate**, the amount of food energy converted per unit time, provides a measure of the total energy used by the body per unit time and can be determined by measuring the heat given off from the body in a defined time period.

4. Factors affecting metabolic rate include:
   a. Exercise.
   b. Ingestion of food.
   c. Hormones such as catecholamines and thyroxine.
   d. Body size, age, and sex.
   e. Fever, disease, emotional state.
   f. Temperature and climate.

5. The metabolic rate of a resting person increases and remains elevated for several hours following ingestion of food.
   a. The specific dynamic action of food varies by food type.
   b. Carbohydrates and fats increase metabolic rate by only 4%.

6. Metabolic rate is roughly proportional to surface area and is commonly expressed as a function of body surface area (Cal/m^2/hr).

7. **Basal metabolic rate (BMR)** is a standard measure for determining metabolic rate that allows comparison between individuals.

8. Measurement of BMR is made under specified conditions:
   a. Fasting at least 12 hours.
   b. In the morning after a night of restful sleep.

c. 30-60 minutes of rest before the test.
d. Comfortable atmospheric temperature about 25°C.

9. A person's metabolic rate can be measured by direct or indirect **calorimetry**.

10. In indirect calorimetry, metabolic rate is based on the determination of a person's oxygen consumption per unit of time, which is measured by a respirometer.
    a. The energy equivalent of oxygen is equal to 4.825 Calories of heat produced per liter of O_2.
    b. Metabolic rate (Cal/m^2/hour) = liters of oxygen consumed times 4.825 Calories per liter of O_2 divided by surface area (m^2).

B. Heat loss from the body involves both physical and physiological processes.

1. The body exchanges heat with its surroundings through four physical processes: radiation, conduction, convection, and evaporation.

2. **Radiation** is heat transfer by electromagnetic radiation between objects not in contact.
    a. Radiant heat exchange is affected by the temperature difference between objects and the body's position.
    b. Temperatures between 21° and 28°C permit the body to radiate heat at approximately the rate of metabolic heat production.
    c. Atmospheric temperature for **thermal comfort** is affected by humidity, wind speed, and level of activity.

3. **Conduction** is heat exchange in the form of kinetic energy between molecules of objects in direct contact.
    a. It is affected by the temperature difference between the objects.
    b. Heat loss or gain by conduction is sizeable in persons exposed to cold water.

4. **Convection** is the transfer of heat between the body and a moving gas or liquid.
    a. Conduction is part of the convective process.
    b. Forced convection increases the rate of heat exchange by the body.
    c. The cooling effect of wind is approximately proportional to the square root of the wind's velocity and accounts for the **"wind chill factor."**

5. **Evaporation** is the process of converting water from a liquid to a gas.
    a. Thermal energy is required to transform water to a gas.
    b. Evaporation, a passive process, is a continuous heat loss mechanism.
    c. This type of heat loss is small and is referred to as heat loss due to **insensible perspiration.**

6. **Sweating** involves active secretion of water by specialized glands in the skin.
    a. It is activated by sympathetic nerves and increases linearly with abnormal increases in body temperature.
    b. Evaporation of sweat is the only way to lose heat in hot environments.
    c. The evaporative capacity of the body decreases with increasing humidity.

C. Heat transfer within the body is essential to maintain a relatively constant body temperature.

1. The transfer of heat within the body is accomplished by a combination of internal conduction and circulatory convection.

2. Heat transfer by convection is slow because of the small temperature difference between adjacent cells and is not adequate to maintain a balance between heat production and loss that is needed to maintain a constant body temperature.

3. Forced convection by the circulatory system is important for thermal balance.
    a. At comfortable temperatures, vasomotor responses regulate peripheral blood flow to achieve thermal balance without sweating or changing metabolic heat production.
    b. Blood flow to the skin determines the rate of heat transfer by circulatory convection from the body core to the skin.

II. Normal Body Temperature

A. Core body temperature is variable in healthy, normal individuals.

1. Body temperature is regulated within a relatively narrow range to ensure the capacity for rapid metabolism.

2. Fluctuations in body temperature produced by changes in atmospheric temperature are

greater at the body surface than in the "homeothermic" core, which ranges from 36° to 39°C.

3. The human body has both longitudinal and radial temperature gradients.

4. Normal body temperature should be conceived as a range of core temperatures that may be expected in a group of normal subjects under various conditions.
   a. Core temperature exhibits temperature gradients from 0.2° to 1.2°C.
   b. The lower and upper limits to the range are 35 and 41°C.
   c. Rectal temperature can be regarded as representative of the body core or "normal" body temperature.

5. Skin temperature is representative of the temperature of the body shell and varies at different body sites and at different atmospheric temperatures.

6. **Mean skin temperature** is regarded as representative of the temperature of the body shell and is calculated as the mean of the temperature at several representative body sites.

> Mean skin temperature = 0.3 (chest and arm temperature) + 0.2 (thigh and calf temperature)

   a. Mean skin temperature is about 33° to 34°C.
   b. In thermography, an infrared camera is used to measure infrared rays emitted from the skin, which are directly related to the temperature of the skin.

B. Mean body temperature is estimated by the fractional sum of the temperatures of the body core and shell.

1. The core temperature, represented by rectal temperature, and the shell, often represented by the mean skin temperature, is estimated from the equation:

> Mean body temperature = 0.3 times mean skin temperature + 0.7 times rectal temperature

   a. The temperatures are taken in a comfortable atmosphere.
   b. The concept of mean body temperature allows for an estimation of the amount of heat gained or lost from the body when exposed to hot or cold environmental conditions.

C. Factors affecting body temperature are variable and multiple.

1. Exercise can increase the core temperature by 2°C or more but decrease the skin temperature because of increased sweat gland activity.

2. Body temperature cycles with a circadian or diurnal rhythm and with fluctuations in temperature averaging 1°C.

3. Changes in temperature associated with the menstrual cycle are superimposed on the diurnal temperature changes.
   a. Temperature drops about 0.6°C before ovulation, after which it rises quickly about 1°C and remains at the normal level until the onset of the next cycle.
   b. The magnitude of oscillations in body temperature varies with the individual.

4. Body temperature in premature infants and the elderly tends to fluctuate with atmospheric temperature.

III. Regulation of Body Temperature

A. Thermoregulation is an example of precise biological negative feedback control and homeostasis.

1. The control of body temperature depends on the coordinated activities of several subsystems, including:
   a. Sweating.
   b. Vasomotor control.
   c. Metabolic rate.

2. The **physiological control system** that coordinates the responses for effective regulation of body temperature is a negative feed back system and has at least three components:
   a. **Temperature receptors** to sense existing body temperature.
   b. **Effector organ systems** to determine heat production and loss.
   c. An **integrator or controller** to compare

sensed temperature to a reference temperature and to activate appropriate effector systems to return temperature to normal.

3. **Peripheral thermoreceptors** are located just beneath the skin. Cold receptors are more numerous than warm receptors.

4. **Central thermoreceptors,** located in the hypothalamus, spinal cord, abdominal viscera, and great veins, include both cold and warm receptors.

5. Nerve impulses from central thermoreceptors are integrated with those from peripheral thermoreceptors in the brain.

6. The hypothalamus, the brain center for body temperature regulation, acts as a thermostat to prevent excessive body heating or cooling.
   a. Stimulation of the hypothalamus by overheating initiates **antirise responses,** such as vasodilation and sweating.
   b. Stimulation by excessive cooling stimulates **antidrop responses,** such as vasoconstriction and shivering.
   c. Somatic and autonomic portions of the nervous system relay hypothalamic signals to effector organ systems.

B. Diseases that affect temperature regulation can result in either regulated or unregulated changes in body temperature.

1. **Fever** is usually caused by **pyrogens,** which appear to reset the hypothalamic thermostat at a higher temperature level.

2. The hypothalamus activates heat production and conservation responses and inhibits heat loss responses as long as the pyrogen is present.

3. **Heat stroke,** a form of heat illness associated with a breakdown of the body's temperature regulating system, may occur if body temperature rises beyond 41° to 42°C.

4. **Heat exhaustion,** a less serious heat illness, is caused by circulatory insufficiency and hypotension.

5. **Cold illnesses** usually involve peripheral circulatory or cardiac defects.

Key Terms

antidrop response
antirise response
basal metabolic rate (BMR)
circadian rhythm
conduction
convection
core temperature
evaporation
heat exhaustion
heat stroke
homeostasis
homeotherm
human calorimeter
hypothermia
hypothalamus
hyperthermia
malignant hyperthermia
mean body temperature
mean skin temperature
metabolic heat
metabolic rate
negative feedback
poikilotherm
pyrogen
radiation
rectal temperature
respirometer
shell temperature
thermography
thermoreceptors

Testing Your Knowledge

I. Matching. Select the correct term from the Key Terms.

________ 1. animal whose body temperature fluctuates with changes in ambient temperature

________ 2. temperature regarded as representative of core temperature

________ 3. representative temperature of body shell

________ 4. technique of determining mean skin temperature by measuring infrared radiation

________ 5. equals 0.3 times mean skin temperature plus 0.7 times rectal temperature

________ 6. amount of food energy converted per unit of time

________ 7. maintenance of normal body temperature

________ 8. insulated chamber for measuring metabolic rate

________ 9. measures oxygen consumption

________ 10. heat transfer by electromagnetic radiation

_________ 11. process of converting water to a gas

_________ 12. heat transfer between the body and a moving gas or liquid

_________ 13. vasodilation and sweating

_________ 14. caused by circulatory insufficiency and hypotension

_________ 15. drop in body temperature to abnormal level

II. Dichotomous. Circle the correct choice from the paired terms in parentheses.

1. (Homeotherms/Poikilotherms) maintain a fairly constant core body temperature.

2. The specific dynamic action of a meal of spaghetti, potatoes, and corn is (greater/less) than a meal of pork chops, potatoes, and corn.

3. A 50-kg 16-year-old girl standing in a 21°C room loses (more/less) heat by (radiation/conduction) than a 50-kg 16-year-old girl sitting in the same room.

4. In an environmental temperature about 20°C, blood vessels in the skin would be (dilated/constricted) to (increase/decrease) heat loss by (radiation/conduction).

5. On a sunny day of 39°C, the temperature of the body shell would be (greater/less) than that of the core.

6. (Rectal/Sublingual/Axillary) temperature most closely represents core body temperature.

7. During strenuous exercise, the (core/shell) temperature is greater than the (core/shell) temperature.

8. In a healthy, fertile, 25-year-old woman, the lowest body temperature would occur just (before/after) ovulation.

9. (Peripheral/Central) thermoreceptors respond to changes in the core temperature.

10. An electrolyte imbalance and absence of sweating may occur during (hypothermia/hyperthermia).

III. True/False. In the space to the left of the number write T for true or F for false.

_______ 1. More chemical energy of food is released as heat in a resting person than someone using a computer.

_______ 2. Maximal secretion of thyroxine has a greater long-term effect on metabolic rate than maximal secretion of epinephrine.

_______ 3. The average metabolic rate of healthy, normal 26-year-old men is about the same as that for 17-year-old women.

_______ 4. A person sitting in a room of 20°C would have a lower basal metabolic rate than a person sleeping in a roomof 20°C.

_______ 5. Core temperature varies over a wider range than shell temperature.

_______ 6. The surface area of the chest and arm is greater than that of the thigh and calf.

_______ 7. In the equation for mean body temperature, the weighing factors for mean skin and rectal temperatures do not change.

_______ 8. Body temperature varies only with activity level during a 24-hour period.

_______ 9. The steady-state discharge rate of warm receptors increases as skin temperature rises but decreases as temperature falls.

_______ 10. The (somatic/autonomic/hypothalamic) part of the nervous system initiates antirise and antidrop responses to overheating or excessive cooling.

IV. Sentence Completion. Write your response in the space provided.

1. Metabolic rate is expressed as __________.

2. The metabolic rate of a 70-kg healthy 25-year-old woman living in Montreal is __________ than the metabolic rate of a

70-kg healthy 25-year-old woman living in San Juan.

3. Two people fall overboard into water 50°F. Person A has 2 m^2 surface area and person B has 1.8 m^2. Heat loss by __________ will be __________ for person A compared to person B.

4. The thickness of the shell is greater in a __________ environment than in a __________ one.

5. The __________ areas of the body appear white on a thermogram.

6. At mean body temperature, about _______ of the body mass is at core temperature and __________ at mean skin temperature.

7. Physiological control of regulation of body temperature involves a _______ _______ system.

8. __________ responses of vasoconstriction and sweating are initiated by the _______ and are delivered to _________________.

9. Fever is a __________ increase in body temperature, while hyperthermia is _______ .

10. Pyrogens affect the hypothalamus indirectly through the actions of __________, __________, and __________.

V. Multiple Choice. Circle the letter of the best response.

1. Which of the macromolecules would be most affected in a person with a fever of 105°F?
   a. DNA
   b. RNA
   c. proteins
   d. carbohydrates
   e. lipids

2. Which person would have the highest metabolic rate?
   a. a man walking at a rate of 3 miles/hour over a hilly path
   b. a person walking at a rate of 3 miles/hour over a hilly path after consuming a meal of pasta
   c. a person walking at a rate of 3 miles/hour over a hilly path after consuming a meal of steak and beans
   d. a person with a thyroxine deficiency walking at a rate of 3.5 miles/hour over a hilly path
   e. a woman walking at a rate of 3 miles/hour over a hilly path after consuming a meal of steak and beans

3. The metabolic rate of a 70-kg 25-year-old man with a surface area of 2 m^2, who consumes 1.6 l of O^2, is:
   a. 38,000 Cal/m^2/hr
   b. 19 Cal/ m^2/hr
   c. 76 Cal/m^2/hr
   d. 12 Cal/m^2/hr
   e. none of the above

4. On a sunny 25°C day, heat loss would be greatest for a person:
   a. walking north at 3 mph with a 5 mph wind from the south
   b. running north at 6 mph with a 5 mph southern wind
   c. walking south at 3 mph with a 5 mph northerly wind
   d. running north at 6 mph with a 5 mph wind from the north
   e. all are equivalent

5. You and a friend are sitting by the pool on a hot summer day. The temperature is 39°C and a 10 mph wind is blowing from the north, the direction you are facing. Your heat loss will occur mainly by:
   a. radiation and convection
   b. conduction and evaporation
   c. convection and evaporation
   d. radiation and sweating
   e. convection and sweating

6. If a person's axillary temperature is 37.4°C, the core temperature is about:
   a. 37.4°C
   b. 36.8°C
   c. 38.1°C
   d. 36.7°C
   e. 38.0°C

7. You are sitting in a 25°C room. The skin temperature of your chest is 36°C and the temperature for your calf, arm, and thigh is 32°C. Your mean skin temperature is:
   a. 33.2°C
   b. 32.8°C
   c. 36.0°C
   d. 37.4°C
   e. 33.5°C

8. Consider the following measurements of temperature in °C: chest skin = 37; skin, arm, thigh, and calf = 33; axillary = 36.7. Your mean body temperature is _______ °C.
   a. 36.4
   b. 34.2
   c. 37.2
   d. 36.7
   e. 38.1

9. The first response to pyrogens released into the body during disease processes is:
   a. secretion of epinephrine
   b. shivering
   c. resting the hypothalamus to a higher set point
   d. vasoconstriction
   e. piloerection

10. Consider the following scenario: It's about 85°F and sunny; a West Point cadet faints while standing at attention for 60 minutes. The most likely reason for fainting is:
    a. heat stroke
    b. heat exhaustion
    c. fever
    d. hypothermia
    e. excessive heat loss by radiation

VI. Complete the Table.

For each condition, indicate the likely effect (increase, decrease, none) effect on the body's temperature.

Condition	Effect on Body temperature
1. Luteal phase of menstrual cycle	
2. Sleep	
3. Exercise	
4. Presence of pyrogens	
5. Ambient temperature of 44 °C	
6. Exposure to cold water	
7. Shivering	
8. Vasodilation and sweating	
9. Heavy work under hot, humid conditions	

Chapter 28

Body Defense and the Immune Response

Chapter Learning Objectives

1. Describe the structure and function of the lymphatic system.
2. Distinguish primary and secondary lymphoid tissues as to function.
3. Describe the various nonspecific defense mechanisms and, in general, how each operates to protect against pathogens.
4. Distinguish the interferons as to function.
5. Understand the basic differences between nonspecific and specific defense mechanisms.
6. Outline the events of cell-mediated immune response to pathogenic bacteria and viruses; describe the types and functions of the chemical signals.
7. List the five classes of immunoglobulins and their functions.
8. Outline the events in antibody-mediated immune response; describe the types and functions of the various molecular signals.
9. Distinguish passive and active immunity and how each develops.
10. Explain suggested mechanisms for autoimmunity.

Expanded Chapter Outline

I. The Lymphatic System

A. Lymph is chemically similar to plasma and interstitial fluid.

1. **Lymph** has a lower protein content than plasma, albumin/globulin (A/G) ratio near 1.0, and high lipid content.

2. Lymph contains all kinds of leukocytes but no erythrocytes.

B. Lymph is drained into systemic venous blood.

1. The three principal types of vessels are capillaries, collecting vessels, and lymphatic ducts.
   a. **Lymphatic capillaries** have a blind end and are more porous than blood capillaries.
   b. Afferent and efferent **lymphatic collecting ducts** carry lymph to and from lymph nodes, respectively, and toward lymphatic ducts.
   c. The connective tissue covering of collecting vessels contains smooth muscle cells and elastic fibers.
   d. Collecting lymphatic ducts have valves.

2. The thoracic and right lymphatic ducts empty into the subclavian veins.

C. Lymphoid tissue contains free and fixed cells.

1. Fixed interconnected reticular cells and fibers form a loose stroma.

2. Five kinds of lymphoid tissue are present in the body:
   a. Diffuse encapsulated lymphoid tissue.
   b. Lymphoid nodules (unencapsulated aggregates of lymphocytes).
   c. Encapsulated lymph nodes (collecting nodules).
   d. Lobes of the thymus gland.
   e. White pulp of the spleen.

D. Primary lymphoid tissues and organs are sites where lymphocytes develop and mature.

1. Bone marrow is the source of red and white blood cells.

2. Each lobule of the **thymus** has a peripheral cortex and central medulla.
   a. Rapidly dividing immature lymphocytes (**thymocytes**) macrophages, and plasma are forced in the cells.
   b. Thymocytes mature into T-lymphatic cells (T-cells) in the thymus.

3. **Peyer's patches** (large aggregates of nodules) are found in the intestinal mucosa.

a. **Germinal centers** in the nodules contain proliferating B-lymphocytes (**plasma cells**), macrophages, dendritic cells, and different kinds of T-lymphocytes.
b. **Antigen binding M cells** are found in the intestinal epithelium covering of Peyer's patches.

E. *Secondary lymphoid tissue and organs are where mature lymphocytes reside and encounter antigen.*

1. Lymph flows one way through cortical and medullary sinuses of lymph nodes.

2. The **lymph node** performs three distinct functions:
   a. Center for lymphocyte proliferation.
   b. Phagocytic organ for purification of lymph.
   c. Site for antibody production and release.

3. The lobules of the **spleen**, the largest lymphoid organ, contain red and white pulp.
   a. Aged erythrocytes are destroyed by resident macrophages in the red pulp.
   b. **Red pulp** is a temporary site of hematopoiesis during fetal development.
   c. The **white pulp** contains T-lymphocytes located adjacent to the central artery and B-lymphocytes in the germinal centers.

II. Internal Defense

A. *Internal defenses comprise mechanisms that deter pathogen invasion of the body and mechanisms that destroy pathogens after invasion.*

1. **Nonspecific defenses** include:
   a. Surface-protective mechanisms.
   b. Groups of defense proteins (interferons, complement).
   c. Defensive processes (inflammation, phagocytosis).

2. **Specific defense** includes mechanisms selectively geared to destruction of specific pathogens and toxins.

B. *Internal defense mechanisms must allow for recognition of self versus nonself.*

1. Self and nonself are distinguished by the types of cell surface markers or antigens.

2. Surface markers may be protein, glycolipid, or glycoprotein.

C. *Nonspecific defenses deter or destroy a variety of pathogens regardless of pathogen type.*

1. **Surface protective mechanisms** include:
   a. Antimicrobial chemicals in sweat, sebum, saliva, tears, and urine.
   b. Commensalic bacteria in skin and mucous membranes.
   c. Skin pH and stomach acid.
   d. Mucous membranes lining body cavities.
   e. Mucus and ciliated epithelial cells in respiratory passages.

2. **Interferons** interfere with viral replication and are released by many types of cells in response to viral infection.
   a. **Interferon alpha** (INF - α) binds to type I receptors and is made by almost all cell types.
   b. **Interferon beta** (INF - β) binds to type I receptors and is made by fibroblasts.
   c. **Interferon gamma** (INF - γ) is produced mainly by activated T-lymphocytes and natural killer cells.
   d. INF - γ binds to type II receptors in membranes and induces target cells to produce proteins that inhibit viral replication, inhibit the growth of tumors, and stimulate specific immune defense.
   e. INF - γ stimulates production of class I or II major histocompatibility complex surface proteins.

3. **Complement**, a series of about 20 proteins that interact in a defined sequence, exhibits four biological actions:
   a. Lysis of cell walls of certain microbes.
   b. Stimulation of histamine release from tissue cells and leukocytes, which leads to an increase in capillary permeability.
   c. Attracting leukocytes (by chemotaxis) to site of reactions.
   d. **Opsonization** and phagocytosis of microbes.

4. **Inflammation**, a localized body response to tissue damage or injury, prevents the pathogen's spread and further tissue destruction.
   a. The clinical signs are redness, heat, edema, and pain.
   b. The sequence of events in inflammation begins with the release of histamine and other chemicals by injured red cells, basophils, mast cells, and others.

5. **Phagocytes** are divided into two groups: neutrophils of polymorphonuclear leukocytes and monocytes/macrophages.
   a. **Neutrophils** are short-lived and arrive first.
   b. Tissue **macrophages** include Kupffer cells (liver), microglia (brain), lung macrophages, others.

6. Macrophages are activated by complement proteins, antigen-antibody complexes, interferon, or other chemicals.
   a. The microbe is engulfed to form a phagosome.
   b. Lysosomes bind to and release digestive enzymes into the phagosome.
   c. Antigenic parts of the microbe may be displayed with **MHC proteins** to activate specific immune defense mechanisms.

D. ***Specific defenses** engage and destroy specific kinds of pathogens: cell-mediated immune response.*

1. Macrophages and **T-lymphocytes** (T-cells) collaborate to destroy viruses and foreign cells entering the body.
   a. Macrophages present the antigen to the T-cell.
   b. Macrophages and T-cells communicate by molecular signals called **cytokines** (interferons, monokines, lymphokines).

2. Maturation of T-lymphocytes occurs in an orderly manner.
   a. Pre-T-cells move through the blood to the medulla of the thymus from the yolk sac, embryonic liver, and bone marrow.
   b. They divide mitotically and acquire cell surface receptors that allow T-cells to be identified and T-cells to identify and distinguish between the body's cells and antigens.
   c. Developing T-lymphocytes possess either a **CD_4** surface marker (**T_4 cells**) and class II MHC proteins or a **CD_8** marker (**T_8 cells**) and **class I MHC proteins**.
   d. Maturing T-cells synthesize cell surface receptors (TCR) for recognition of foreign antigens.
   e. The two peptide chains of the TCR receptor determine its antigen and MHC specificity.
   f. Negative selection removes T-cells whose receptors react with self-MHC and other self-proteins. (About 96-99% of developing T-cells are negatively selected.)
   g. The mature, immunocompetent T-cells reenter the blood to colonize other lymphoid tissues and react to antigens presented in combination with class I MHC or class II MHC.
   h. Each competent T-cell after binding to its designated antigen is stimulated to divide mitotically (**clonal expansion**).

3. Functionally, there are two kinds of CD_8^+ (class I MHC restricted) cells: **cytotoxic (T_c)** and **suppressor (T_s)**; and two kinds of CD_4^+ (class II MHC restricted) cells: **helper** (T_H) and **memory (T_M)**.

4. The sequence of events in **cell-mediated immune response** to a **bacterium** includes:
   a. Ingestion of the bacterium by a macrophage and breakdown of bacterial proteins into **epitopes (antigenic determinants)**.
   b. Combining of epitopes with class II MHC and movement to membrane.
   c. Presentation and binding of MHC-antigen to receptor on competent resting CD_4 cell.
   d. CD_4 activated by macrophage membrane protein, CD80, which binds to CD28 on CD_4 cell.
   e. Stimulation of macrophage to secrete interleukin-1 (IL-1).
   f. IL-1 stimulates $T_4 \rightarrow T_4$ helper, T_H0.
   g. IL-2 stimulates clonal expansion of T_H0 to form memory cells and CD_4+**T_H1** cells.
   h. CD_4+T_H1 cells secrete INF-, which activates macrophage to destroy ingested antigens.

5. The sequence of events in **cell-mediated immune response** against a **virus** is:
   a. Viral entry and synthesis of viral proteins by host cell.
   b. Combining of viral proteins and class I MHC and incorporation into plasma membrane of host cell.
   c. Production of and insertion of CD80 into membrane of antigen-presenting cell.
   d. Activation of resting competent **cytotoxic CD_8^+** cell bound to presented MHC antigen through CD28 of CD_8^+.
   e. Destruction of virally infected cell.

6. **Suppressor T-cells** develop from activated CD_8^+ cells and turn off activated cytotoxic cells when pathogens are no longer a threat.

E. *Antibodies are produced by specialized B-lymphocytes in response to specific antigens.*

1. The basic structure of all **immunoglobulins** is two identical light chains and two identical heavy chains stabilized and linked by disulfide bonds to form a Y-shaped molecule.
   a. The **light chains** exist in two forms: kappa and lambda.
   b. The **heavy chains** exist in five major forms: alpha, gamma, delta, epsilon, and mu, and designate the class of antibody (IgA, IgG, IgD, IgE, and IgM).
   c. Each heavy and light chain has a **constant region** (CH, CL) and a **variable region** (VH, VL).
   d. Constant regions bind complement and initiate complement reactions.
   e. Variable regions of heavy and light chains contain **antigen-specific binding sites** and can bind to two of the same kind of antigen to form antigen-antibody complex.

2. Immunoglobulins (Ig) are grouped into five classes based on the structure of the heavy chain.
   a. **IgG**, the major immunoglobulin, protects against pathogens and toxins.
   b. **IgA**, the principal immunoglobulin in secretions (mucus, saliva, milk, tears, fluids in genitourinary and digestive tracts) protects against pathogens that contact skin, are ingested, or are inhaled.
   c. **IgM** works with complement and phagocytes to protect against bacteria; IgM antibodies are used to characterize ABO blood type.
   d. **IgD**, located in plasma membranes of circulating B-lymphocytes, are believed to play a role in antigen-triggered lymphocyte differentiation (development of plasma cells and memory cells from B-lymphocytes).
   e. **IgE**, located on the surface of basophils and most cells, stimulate these cells to release histamine and other chemicals to mediate a local response to the antigen's presence.
   f. IgE molecules are directly involved in diseases characterized by **hypersensitivity** (asthma, hay fever) and defend against helminthic parasites.

3. An antibody combines with its specific antigen to form an **antigen-antibody complex**.
   a. The complex may stimulate phagocytosis.
   b. The antigen's ability to harm may be neutralized.
   c. The complex may activate complement, resulting in phagocytosis and opsonization of the pathogen.

F. *The antibody-mediated immune response is a defense mechanism designed to destroy specific pathogens.*

1. Competent **B-lymphocytes** recognize and bind to a specific antigen.

2. The sequence of events in antibody-mediated immune response is:
   a. Presentation of bacterial antigen, class II MHC protein and CD80 protein, or macrophage plasma membrane.
   b. Recognition and binding by competent CD_4^+ to antigen-MHC II complex and CD80 via CD28 on CD_4^+ cell.
   c. Stimulation of CD_4^+ cell by interleukins 1 and 4 to develop initially into a helper cell (TH0) and then differentiate further into **T_H2**, a CD_4^+ helper cell that mediates the antibody-mediated immune response.
   d. Binding and displaying antigen with MHC II and CD40 on surface of competent B-lymphocyte.
   e. Activation of competent B-lymphocyte by binding of T_H2 to MHC-II antigen on B-lymphocyte through T_H2 CD40 ligand.
   f. Stimulation by interleukins 4, 5, and 10 of activated B-lymphocyte to differentiate into a clone of **plasma cells and memory B-cells**.
   g. Plasma cells secrete antibodies chemically identical to antigen receptors into lymph and blood.

3. Memory B-cells permit a rapid mobilization on subsequent exposure to the same antigen and form the basis of lasting immunity.

G. *Immunity against pathogens can be induced.*

1. **Active immunity** confers long-lasting, continuing protection.
   a. **Immunization** stimulates the immune system to produce antibodies against the antigen.
   b. The antigen may be an attenuated bacteria or virus (tuberculosis, measles,

mumps), unable to cause disease but able to stimulate an immune response, or an antigen protein produced through recombinant DNA technology (hepatitis B antigen).

2. **Antibody titer** is used as a measure of amount of specific antigen present.
   a. In the primary response, the titer rises and then falls within the first six weeks.
   b. If the antigen is given during the decline, a secondary response occurs with a higher titer and longer-lasting immunity.

3. **Passive immunity** confers temporary, short-term protection.
   a. Antibodies are given to a person to combat disease.
   b. The duration depends on amount of antibody given, frequency of administration, and rate of recipient's metabolism of antibodies.
   c. The antibodies or immunoglobulins are isolated from a donor with active immunity or from an animal (horse) stimulated to produce them.
   d. **Monoclonal antibodies**, produced by hybridomas, are used diagnostically.
   e. Newborns have maternal antibodies acquired during pregnancy and during breast-feeding.

H. *Autoimmunity occurs when lymphocytes and antibodies react with normal components of the body.*

1. Physiological autoimmunity responses are part of the body's daily maintenance of homeostasis.
   a. Senescent erythrocytes are removed from circulation and destroyed in the spleen.
   b. Changes in the membrane expose a binding site for IgG, which, when bound, induces phagocytosis by macrophages.

2. **Autoimmune disease** results from a breakdown in recognition and tolerance of self.
   a. An immune response is directed against body cells and tissues.
   b. The disease may involve many organs (systemic lupus erythematosis) or one (type I diabetes mellitus).

3. Possible inducers of autoimmune disease include:
   a. Loss of suppressor T-cells.
   b. Viral infections, which activate autoreactive T-cells or directly infect B-cells.

4. The development and severity of autoimmune disease is determined in part by the type of MHC molecules possessed by the affected person.
   a. MHC molecules are genetically determined.
   b. Autoimmune disease may result when class II MHC protein allows presentation of self-antigens to activated T-cells.

Key Terms

active immunity	histamine
anaphylaxis	lymphokines
antibody	lysozyme
antibody-mediated immune response	monoclonal antibody
antigen	nonspecific defense
antigen-presenting cell	opsonization
autoimmunity	passive immunity
B-cell	pathogen
bone marrow	phagocytosis
CD_4^+ cell	plasma cell
CD_8^+ cell	specific defense
cell-mediated immunity	T-cell (T_H, T_C, T_S, T_M)
class I MHC	thymus
class II MHC	immunoglobulins
complement	inflammation
constant region	interferons
cytokines	interleukins
	lymph
	lymph node

Testing Your Knowledge

I. Matching. Select the correct term from the Key Terms.

________ 1. encapsulated collections of nodules

________ 2. encapsulated collections of lymphoid cells form lobules in this

________ 3. disease-causing organism

________ 4. foreign material stimulating an immune response

________ 5. protein composed of 2 heavy and 2 light chains

_________ 6. life-threatening allergic reaction

_________ 7. precursor of antibody-forming plasma cell

_________ 8. immune reaction mediated by T-cells

_________ 9. subset of CD_4^+ lymphocytes necessary for antibody-mediated response

_________ 10. molecule promoting T-cell proliferation

_________ 11. molecules secreted by lymphocytes, which have a variety of effects on other cells

_________ 12. cluster of genes encoding cell surface antigens that define self

_________ 13. produced by hybridoma cells

_________ 14. proteins protecting cells against viral pathogens

_________ 15. region of Ig molecules specific for antigen

II. Dichotomous. Circle the correct choice from the paired terms in parentheses.

1. Compared to plasma, lymph has (more/less) protein and (more/less) lipid.

2. (Afferent/Efferent) lymphatic vessels carry lymph to lymph nodes, and (afferent/efferent) vessels carry lymph from nodes to lym phatic ducts.

3. The "flapper" valves in lymphatic capillaries open when colloid osmotic pressure is (greater/less) than hydrostatic pressure.

4. Lymphocytes develop and mature in (primary/secondary) lymphoid tissues and react with antigen in (primary/secondary) lymphoid tissues.

5. Phagocytosis and inflammation are two types of (nonspecific/specific) defense mechanisms.

6. CD_8^+ cells express class (I/II) MHC molecules.

7. (B-/T-) lymphocytes produce and secrete antibodies in response to specific antigens.

8. The principal antibody in body secretion is (IgA/IgD/IgE/IgG/IgM).

9. Memory cells are essential elements in (passive/active) immunity.

10. The HIV virus infects and eventually destroys (TH/TM/Tc/Ts) lymphocytes.

III. True/False. In the space to the left of the number write T for true or F for false.

_______ 1. The thymus is the same size in a 20-year-old woman and a 50-year-old man.

_______ 2. Bone marrow is classified as both a primary and a secondary lymphoid tissue.

_______ 3. B- and T-lymphocytes are released directly into blood flowing through lymph nodes.

_______ 4. Commensal bacteria cannot become pathogens.

_______ 5. Only certain leukocytes can become macrophages or phagocytic cells.

_______ 6. TS and TM cells are identified as CD_4^+ class II MHC restricted cells.

_______ 7. CD28 is found on both CD_4^+ and CD_8^+ cells.

_______ 8. B-lymphocytes develop competency in the thymus.

_______ 9. An example of active immunity is the transferring of antibodies from mother to fetus during pregnancy and to the infant during breast-feeding.

_______ 10. IgE plays a role in allergic reactions classified as type I hypersensitivity.

IV. Sentence Completion. Write your response in the space provided.

1. The thoracic and right lymphatic ducts empty lymph into the _____________.

2. Erythrocytes and leukocytes are produced in ____________ after birth but in the ____________ during embryonic and fetal stages of development.

3. During inflammation, ____________ increases the permeability of capillaries.

4. Macrophages and lymphocytes communicate by molecular signals called ____________ .

5. Macrophages and T-lymphocytes are important cellular components of ____________ .

6. Immunocompetent T-lymphocytes develop and mature in the ____________ .

7. CD80, produced by a(n) ____________ binds to CD28 on the surface of a(n) ____________ .

8. ____________ is necessary for the production of T_M and T_H1 from T_H0 and the differentiation of T_S and T_C from CD_8^+ cells.

9. ____________ lymphocytes destroy virally infected cells presenting class I MHC^+ antigen.

10. Each immunoglobulin can bind to ________ antigen(s) through the __________ regions of the heavy and light chains.

V. Multiple Choice. Circle the letter of the best response.

1. The functions of lymph nodes include all of the following except:
   a. lymphatic proliferation
   b. B-cell maturation
   c. T-cell maturation
   d. purification of lymph
   e. antibody production and release

2. Germinal centers for B-lymphocytes are located in:
   a. the spleen
   b. Peyer's patches
   c. lymphoid nodules
   d. lymph nodes
   e. all of the above

3. Which of the following molecules is not part of a nonspecific defense mechanism?
   a. histamine
   b. lysozyme
   c. interferon
   d. immunoglobulin
   e. hydrochloric acid

4. Which of the following does not describe interferon gamma?
   a. produced mainly by activated T-lymphocytes
   b. binds to type II receptors
   c. induces macrophages to kill virally infected cells
   d. stimulates production of class I or II MHC proteins
   e. used to treat genital warts, Kaposi's sarcoma, and hairy cell leukemia

5. The biological actions of complement include all the following except:
   a. lysis of cell walls of certain microbes
   b. action as chemotaxins
   c. stimulation of histamine release
   d. preparation of microbes for opsonization
   e. vasodilation

6. The lymphocyte essential for both cell-mediated and antibody-mediated immune responses is the:
   a. CD_4^+
   b. CD_8^+
   c. TC
   d. TH
   e. TM

7. Which of the following events occurs first in T-cell activation?
   a. secretion of cytokines
   b. induction of mitotic activity
   c. expression of new cell surface antigens
   d. early signal transduction events
   e. destruction of pathogen

8. The immunoglobulin involved in hypersensitivity reactions and protection against parasites is:
   a. IgA
   b. IgD
   c. IgE
   d. IgM
   e. IgG

9. An activated B-lymphocyte differentiates into a plasma cell as a result of:
   a. stimulation by IL-4
   b. binding to a T_H2 through CD40
   c. stimulation by IL-4, 5, and 10
   d. binding to antigen-presenting

macrophage
e. b and c are correct

10. Possible causes and inducers of autoimmune disease include all the following except:
    a. alteration of self-antigen so it is recognized as foreign
    b. loss of T_S cells
    c. infection with microbe or virus carrying a cross-reactive antigen
    d. absence of thymus
    e. type of MHC molecules

Chapter 29

Environmental Physiology

Chapter Learning Objectives

1. Tell how the body reacts to high altitudes.
2. Describe the different conditions that develop at high altitudes.
3. Give the physiological effects of prolonged exposure to high altitudes.
4. Summarize the respiratory phenomena associated with deep sea diving.
5. Outline the problems related to too much oxygen.
6. Explain how the presence of other gases affects respiration.
7. Indicate how the phenomenon of weightlessness affects breathing.
8. Describe the principal effects of extremes in temperature on health.
9. Describe the principal effects of respiratory and systemic pollutants on health.

Expanded Chapter Outline

I. Response to the Environment

A. Environmental stresses cause adaptation.

1. The integrated nervous and endocrine systems stimulate homeostatic adjustments in response to environmental stresses.

2. Immediate and long-term physiological adjustments may involve homeostasis, adaptation, acclimation, and acclimatization.

B. Environmentally induced changes lead to acclimatization.

1. **Acclimation** is a functional change occurring in response to a single environmental stress that is artificially induced in the laboratory.

2. **Acclimatization** refers to physiological changes occurring in response to natural environmental stresses.

II. Biological Clocks

A. Biological rhythms have intrinsic clocks.

1. These **biological rhythms** may occur at the cellular level (e.g., pacemaker cell) and the whole body level (e.g., metabolic rate, temperature).

2. They vary in time from seconds to minutes to days.

3. They are often neuronal in origin.

*B. **Circadian rhythms** track the **light-dark cycles** (L-D cycles).*

1. Physiological responses operating in the L-D cycles include metabolic rate, temperature, heart rate, and hormone secretion (e.g., melatonin, cortisol).

2. Many rhythms persist if L-D cycles are altered.

3. Seasonal affective disorder (SAD) is linked to lack of adequate daylight.

4. The hypothalamus and pineal gland are involved with circadian rhythm.
   a. In mammals, a central biological clock is believed to be located in the suprachiasmatic nucleus.
   b. The eye and **pineal gland** have a neural connection to the suprachiasmatic nucleus.

5. The lunar cycle with a period of 29.5 days has little effect on biological rhythms in humans.

III. Responses to the Physical Environment

*A. High altitude causes **hypoxia**.*

1. There is an inverse relationship between

altitude and atmospheric pressure, alveoli gas tension, and arterial O_2 saturation.

2. Physical and physiological reposes to increasing altitude include:
   a. **Hyperventilation** (deeper breaths taken more rapidly).
   b. Increased cardiac output to match ventilation.
   c. Increased excretion of bicarbonate by kidney.

3. The sequence of events for hyperventilation is:

> fall in P_{IO2} → decrease in P_{AO2} → decreased arterial oxygen tension → increased firing carotid bodies → hyperventilation → increased alveolar and arterial oxygenation

4. The loop is regulated by a negative feedback limb that adjusts for decreased arterial CO_2 tension resulting from hyperventilation.

5. Because of acid-base disturbances, hypoxic hyperventilation occurs in two stages: (1) immediate increase in ventilation and (2) slowly rising increase in ventilation over 8 to 12 hours with sustained ventilation, often 12 hours.
   a. The initial rapid increase in ventilation causes respiratory alkalosis.
   b. The increased pH depresses respiration.
   c. Renal compensation, which occurs within 10 to 12 hours, restores the acid-base balance.

6. Long-term physical and physiological responses to low environmental oxygen include:
   a. Increased production of erythrocytes.
   b. Increased ventilation, cardiac output, and circulatory hemoglobin, which tend to return to low altitude values over a long period.
   c. Cellular changes in the number of mitochondria and alterations of metabolic pathways.

B. *Korean women are champion divers.*

1. The divers, called Ama, descend to depths of 80 feet and hold their breaths for up to 2 minutes.
   a. Divers hyperventilate for 5 to 10 minutes before diving.
   b. They whistle with each forced expiration to prevent too much CO_2 from being blown off.
   c. They do not take a full breath before diving to extend the depth of the dive.

2. Effects of diving on blood gases include:
   a. Decrease in alveolar O_2 concentration, but increase in P_{AO2} because of lung compression.
   b. Decrease in alveolar CO_2, but increase in P_{AO2} because of lung compression.

3. During ascent, the lungs expand and reverse the situation created by compression.
   a. CO_2 rapidly diffuses from blood to lungs.
   b. O_2 content of arterial blood drops because P_{AO2} is similar to PO_2 in pulmonary capillary bed.

4. Physiological changes occurring in the "**diving reflex**" include:
   a. Vasoconstriction except in blood vessels to the heart, brain, and exercising muscles.
   b. Decreased heart rate.

5. **Acclimatization** to diving may be accomplished through training and includes:
   a. Increasing lung vital capacity.
   b. Elevation of basal metabolic rate.

C. *Increased oxygen pressure can be lethal.*

1. Pressure of water is directly proportional to depth because of the weight of water.
   a. Pressure increases by 1 atmosphere for every 33 feet below the surface.
   b. At 33 feet underwater, the alveoli pressure is 2 atmospheres.

2. A characteristic of devices used for diving is that air enters the lungs at higher than normal pressure, and significant amounts of nitrogen in the compressed air are absorbed into the tissue.
   a. More nitrogen is dissolved in a longer dive.
   b. Nitrogen is released from tissues during ascent.
   c. If ascent is slow, nitrogen can be picked up by the blood and returned to the lungs to be exhaled.
   d. With rapid ascent, bubbles form in the

blood and tissues, a syndrome known as the "**bends**," which is treated in a decompression chamber.

e. At depths below 200 feet, nitrogen has a narcotic effect that may be avoided by switching from a N-O_2 mixture to a He-O_2 mixture.

3. Adverse effects of breathing pure oxygen include:
   a. **Atelectasis** (collapsed alveoli), which occurs in pilots, astronauts, and hospital patients when small airways become blocked by mucus.
   b. **Toxicity**, which primarily affects the lung and can cause pulmonary edema.
   c. **Retrolental fibrosis** in premature infants.
   d. Convulsions when O_2 is breathed at pressures exceeding 760 mm Hg.

4. The first site of injury of lung oxygen toxicity is the endothelial cells of the pulmonary capillaries.
   a. Free radicals, by-products of molecular oxygen, are rapidly removed by protective enzymes (superoxide dismutase, catalase, peroxidase) under normal conditions.
   b. At high oxygen concentrations, excess free radicals cause destructive and lethal cellular effects by oxidizing membrane lipids, cellular proteins and DNA, and enzymes, thus shutting down cellular respiration.
   c. Free radical-induced injury to the lung may result from breathing ozone or NO_2 or exposure to paraquat.

5. Retrolental fibrosis causes blindness in premature infants.
   a. The mechanism for blindness is deterioration of the retina and inadequate development of retinal blood vessels.
   b. Retrolental fibrosis may be avoided by keeping PO_2 less than 140 mm Hg in the incubator.

D. Increased carbon dioxide and carbon monoxide pressures can be lethal.

1. Rebreathing higher than normal levels of CO_2 stimulates ventilation.
   a. 80 mm Hg P_{CO2} is the maximum tolerated concentration.
   b. At levels greater than 80 mm Hg, inhibition of respiratory centers occurs, respiration fails, and acidosis becomes severe, leading to lethargy and unconsciousness.

2. Carbon monoxide (CO) does not stimulate respiration.
   a. CO competes with O_2 for the same binding sites on hemoglobin.

   Hb + CO $\Leftrightarrow$ HbCO (carboxyhemoglobin)

   b. The reaction is a function of P_{CO}.
   c. The binding affinity of CO for Hb is about 210 times that of O_2.
   d. A P_{CO} of 0.7 mm Hg (0.1% of air) can be lethal.
   e. The brain is the first organ to show sensitivity to anoxia.

3. CO poisoning is not detected by the body.
   a. There are no chemoreceptors for CO.
   b. Chemoreceptors that detect changes in P_{O2} are not stimulated because the blood P_{O2} is normal.

4. Treatment for CO poisoning is administration of 100% O_2 and 5% CO_2.

E. Space travel imposes unusual stresses.

1. The human body's sensation of weight is due to gravitational attraction and is referred to as **G force**.
   a. The G force expresses the magnitude of accelerating forces.
   b. G forces may be positive or negative.

2. With a positive G force exerted in a head to foot direction, blood is pulled toward the lower parts of the body.
   a. At G forces greater than or equal to 4, blood pools in lower extremities, arterial systemic pressure falls, blood flow to the brain is reduced, and blackout may occur.
   b. The effects experienced by astronauts at take-off and landing are minimized by wearing antigravity suits and being in a reclined position.

3. With negative G force (foot-head direction), blood is pulled toward the head.
   a. At or above negative 3 G, neck veins are overdistended, vision is impaired, and cerebral hemorrhage may occur at higher negative G.

b. Physiological effects of negative G are much longer lasting than those of positive G.

4. The physiological problems associated with weightlessness include:
   a. Lack of hydrostatic pressure, resulting in decreased blood volume, red cell mass, and cardiac output.
   b. Diminished physical activity, which leads to decreased work capacity and loss of calcium from bones.

F. Temperature extremes cause injury.

1. Physiological responses to cold stress include:
   a. Cutaneous vasoconstriction.
   b. Piloerection (goose bumps).
   c. Neurohormonal activation of the hypothalamus.
   d. Increased electrical activity of skeletal muscles (shivering).
   e. Changing body position to reduce heat loss.
   f. Increased heart rate and ventilation.

2. Most of the responses are initiated through sympathetic stimulation, which tends to conserve heat.

3. Acclimatization to cold occurs over a two- to three-week period and involves:
   a. Switch from shivering to nonshivering **thermogenesis**.
   b. Increase in food intake.
   c. Elevation of metabolic rate due to level of thyroid hormones.
   d. Rise in skin temperature, warmer hands and feet, and higher core temperature.

4. **Heat stress** is usually classified as hot-day or hot-humid.
   a. Hot-day is characterized by a temperature greater than 95°F and low humidity.
   b. Hot-humid is characterized by an upper limit for ambient temperature of 90° to 95°F and relative humidity greater than 75%.

5. Short-term physiological responses to heat stress are:
   a. Increased sweating and cutaneous vasodilation.
   b. Transitory increase in plasma volume with shift of fluid from interstitial space to vascular bed.
   c. Decreased blood volume.
   d. Increased O_2 consumption because of elevated temperature and metabolic rate.
   e. Increase in food consumption through body weight decrease.
   f. Dehydration.

6. Acclimatization to heat stress takes 7 to 14 days. Specific changes are:
   a. Increased peripheral thermal conductance.
   b. Increased sweating capacity with fall in salt.
   c. Rise in skin temperature, fall in core temperature, and more even distribution of sweat over the body.

7. Failure to tolerate heat stress leads to **heat exhaustion**.
   a. Heat exhaustion is characterized by subnormal body temperature, dizziness, headache, nausea, and sometimes collapse.
   b. The first stage of heat exhaustion, termed **heat shock**, may be characterized by a critical rise in body temperature and heat-induced tetany (muscle cramps and uncontrolled muscle spasms).
   c. The second stage of heat exhaustion, termed **heat stroke** (sunstroke), is marked by convulsions, coma, and brain damage.
   d. Massive vasodilation and decrease in blood volume cause circulatory shock.

IV. Response to Chemical Environment

A. Respiratory pollutants exhibit different modes of action.

1. Three important features of **chemical pollutants** are mode of entry, removal, and tissue storage.
   a. Chemicals gain entry through the lungs, gastrointestinal tract, and skin.
   b. Chemicals are removed by excretion, defecation.
   c. Chemicals may be detoxified or transformed in the liver before elimination.

2. Pollutants are classified as respiratory or systemic.

3. Respiratory pollutants exhibit different modes of action and affect different types of lung cells.

4. The major sulfur pollutants, SO_2, H_2S,

H_2SO_4, and SO_4^{-2}, cause swelling of membranes, increased mucous secretion, narrowing of airway passages, and alveolar thickening.

5. **Oxides of nitrogen**, NO_2, NO, attack biological membranes, and exposure may result in pulmonary edema and lung damage.

6. **Ozone** at high concentrations can cause bronchitis and emphysema and may be involved in premature aging and transformation of lung tumors.

7. Respiratory dusts (particles of carbon, quartz, silica, sulfur, cobalt, and iron oxide) can cause **pneumoconiosis**, a condition characterized by **fibrosis** (scarring of lung tissue).
   a. Respiratory dust particles range from 0.5 to 1.5 μm.
   b. Particles from 5 to 10 μm are stopped by hairs in the nose and moist membranes in nasopharangeal passages.
   c. Particles from 2 to 5 μm are trapped in mucus in the tracheobronchial tree.
   d. Particles less than 2 μm may reach the alveolar/capillary bed, where alveolar macrophages may remove them.
   e. Unremoved particles may cause injury and pneumoconiosis, silicosis, or asbestosis.

8. **Pneumoconiosis**, regardless of its cause, has similar pathophysiology.
   a. The first stage leads to scarring of lung tissue (fibrosis), loss of lung stiffness, shortness of breath, and decline in pulmonary function (difficulty exhausting air from lungs).
   b. The latter stage leads to emphysema and **hypoxemia**.
   c. Hypoxemia triggers vasoconstriction of pulmonary arterial vessels, causing pulmonary hypertension and eventually right heart failure.

B. *Systemic pollutants (lead, mercury, fluoride, and cadmium) affect numerous organs and may accumulate in the body.*

1. Lead, which enters through the lungs and gastrointestinal tract, has four sites of action: blood, gastrointestinal tract, nervous system, and kidney.
   a. Early symptoms are anemia, headaches, and muscle pain.
   b. Later symptoms are vomiting, abdominal pain, black stools.
   c. Advanced stages include brain damage, convulsions, renal failure.

2. Mercury, which enters the body as an organic or nonorganic form primarily through the food chain, can damage several organs (kidney, brain, and liver).

3. Cadmium enters the body primarily through food and water and through the lungs and concentrates in the kidney, lungs, and liver.
   a. Long-term cadmium poisoning leads to pneumonia, pulmonary emphysema, and cirrhosis of the liver and is also linked to hypertension.
   b. Cadmium acts on mitochondrial enzymes to block ATP synthesis and inhibits ATPase to prevent the body from using ATP.

Key Terms

acclimation
acclimatization
adaptation
anoxia
asbestosis
bends
biological rhythms
blackout
carboxyhemoglobin
circadian rhythms
diving reflex
free radicals
G force
heat shock
heat stroke
hyperventilation
hypoxemia
hypoxia
oxygen radicals
oxygen toxicity
pneumoconiosis
polycythemia
radiation
silicosis
weightlessness

Testing Your Knowledge

I. Matching. Select the correct term from the Key Terms.

________ 1. results from hematocrit > 70%

________ 2. physiological change occurring in response to single induced environmental stress

________ 3. physiological changes occurring in response to natural environmental stresses

________ 4. condition of low oxygen

_________ 5. deeper, more rapid breathing

_________ 6. formed from binding of CO and Hb

_________ 7. heat exhaustion stage marked by convulsion, coma, and brain damage

_________ 8. condition characterized by fibrosis

_________ 9. condition of collapsed alveoli from breathing pure O_2

_________ 10. condition of oxygen deprivation

_________ 11. condition of low arterial Po_2

_________ 12. oxygen by-product

II. Dichotomous. Circle the correct choice from the paired terms in parentheses.

1. (Adaptation/Acclimatization) results from a genetic change.
2. The stimulus for ventilation during hypoxia is a decrease in (arterial Po_2/% O_2 saturation).
3. The increase in erythrocytes that occurs in a person living at 15,000 feet altitude is an example of (acclimation/acclimatization).
4. During a 40-foot dive, alveolar partial pressure of oxygen in the lungs of Ama divers (increases/decreases) compared to before the dive.
5. At the end of hyperventilation and before the Ama dive, the alveolar O_2 concentration is (greater/less) than the value before hyperventilation.
6. Hemoglobin has a (higher/lower) affinity for O_2 than for CO.
7. During takeoff, a person is subjected to acceleration forces moving in a (head-to-toe/toe-to-head) direction and (positive/negative) G forces.
8. A decrease in blood volume may occur in response to (cold/heat).
9. (Cadmium/Lead/Mercury) interferes with utilization of ATP as a source of energy.
10. Cobalt and asbestos are classified as (systemic/respiratory) pollutants.

III. True/False. In the space to the left of the number write T for true or F for false.

_________ 1. Circadian rhythms are always tied to the light-dark cycle.

_________ 2. The percent of oxygen in the air at 12,000 feet above sea level is lower than at 2000 feet.

_________ 3. Polycythemia is characterized by a hematocrit less than 70%.

_________ 4. At a depth of 40 feet, alveolar CO_2 in the lungs of Ama divers increases compared to before the dive.

_________ 5. At a depth of 300 feet, a seal's heart rate decreases.

_________ 6. The affinity of hemoglobin for oxygen is greater than for CO_2.

_________ 7. Women do not acclimatize to heat as well as men.

_________ 8. More lead enters the body by way of the lungs than the digestive tract.

_________ 9. A common site of action for lead, cadmium, and inorganic mercury is the kidney.

_________ 10. Mercury enters the body primarily through drinking and breathing.

IV. Sentence Completion. Write your response in the space provided.

1. The central biological clock is believed to be located in the ___________.
2. To correct the respiratory alkalosis that occurs in the first stage of hypoxic hyperventilation, the kidney excretes __________.
3. In the lungs of Ama divers, at a depth of 40 feet CO_2 moves from ________ to ________; during the ascent, CO_2 moves from ________ to ________.
4. In the lungs of Ama divers, at a depth of 40 feet, O_2 moves from ________ to ________;

during the ascent, O_2 may move from _______ to _______.

5. The maximum tolerable alveolar Pco_2 is _________ mm Hg.

6. Consider the reaction: Hb + CO ⇔ HbCO. At a PCO of 0.5 mm Hg, the reaction shifts to the _________.

7. _________ thermogenesis is one of the first changes to occur during acclimatization to cold.

8. Particles _________ μm may reach the alveolar/capillary bed, where they are removed by _________.

9. A common source of lead in a house is _________.

10. Chlorinated hydrocarbons like DDT accumulate in _________ and _________ tissues.

V. Multiple Choice. Circle the letter of the best response.

1. All of the following exhibit a circadian rhythm even in the absence of light except:
   a. heart rate
   b. metabolic rate
   c. milk production
   d. temperature
   e. cortisol secretion

2. As altitude increases from sea level:
   a. baropressure increases
   b. the percent O_2 decreases
   c. alveolar oxygen pressure increases
   d. alveolar CO_2 pressure increases
   e. arterial O_2 saturation (in percent) decreases

3. Which of the following events occurs first as the body responds to altitude by hyperventilation?
   a. decreased arterial O_2 tension
   b. decreased alveolar O_2 tension (P_{AO2})
   c. fall in inspired O_2 tension (P_{IO2})
   d. increased firing of carotid bodies
   e. decreased arterial CO_2 tension

4. After one year at an altitude of 14,000 feet, all of the following physiological parameters tend to return to low-altitude values except:
   a. cardiac output
   b. ventilation
   c. circulating hemoglobin
   d. work capacity
   e. none of the above

5. Compared to predive conditions, all of the following physiological changes occur in the lungs of Ama divers at a depth of 40 feet except:
   a. decrease in alveolar O_2 concentration
   b. increase in P_{AO2}
   c. decrease in alveolar CO_2 concentration
   d. decrease in P_{ACO2}
   e. compression of lungs

6. All of the following pathological conditions result from oxygen toxicity except:
   a. atelectasis
   b. bends
   c. retrolental fibrosis
   d. pulmonary edema
   e. convulsions

7. Under a condition of weightlessness:
   a. blood pools in the legs
   b. a person may black out
   c. redout may occur
   d. retinal hemorrhage may occur
   e. cardiac output decreases

8. A response common to both cold stress and heat stress is:
   a. cutaneous vasodilation
   b. increased metabolism
   c. increased heart rate
   d. increased ventilation
   e. dehydration

9. Pulmonary edema may occur with inhalation of:
   a. H_2SO_4 mist
   b. SO_2
   c. NO_2
   d. CO_2
   e. O_3

10. Pneumoconiosis may be caused by all of the following except:
    a. silica
    b. asbestosis
    c. cobalt
    d. iron oxide
    e. lead

Chapter 30

Physiology of Exercise

Chapter Learning Objectives

1. Explain where the energy comes from to allow muscle contraction; describe the three metabolic systems that supply energy for muscle contraction.
2. Compare the glycogen-lactic system with the aerobic energy system.
3. Indicate what happens to oxygen consumption during exercise and what happens in the build-up of recovery oxygen; draw a graph showing oxygen uptake during increasing levels of steady-state exercise.
4. Describe the utilization of nutrients during exercise, the relationship between diet and availability of nutrient, and the rate of replenishment as it is affected by diet.
5. Describe ventilation during exercise.
6. Detail the stimulation of pulmonary stimulation during exercise by increased carbon dioxide concentration, increased pH, and decreased oxygen concentration.
7. Indicate how the nervous system affects ventilation during exercise.
8. Indicate what happens to blood flow during exercise.
9. Compare untrained and trained individuals at rest and during exercise as to heart rate, cardiac output, and stroke volume.
10. Describe how the body regulates temperature during exercise.

Expanded Chapter Outline

I. Overview

A. Exercise is often classified as either dynamic or static.

1. **Dynamic exercise** (stair climbing, bicycling, running) is performed anytime muscle length changes.
   a. In **positive** dynamic exercise, the muscles act as a "motor."
   b. In **negative** dynamic exercise, the muscles act as "brakes."

2. **Static exercise** is performed during isometric muscle contraction.

3. Elevated metabolic demand is the common feature of all types of exercise.

II. Metabolic Aspects of Exercise

A. Energy requirements of exercise are provided by high-energy phosphate compounds.

1. Energy is provided by three metabolic systems:
   a. The phosphagen system.
   b. The glycogen-lactic acid system.
   c. The aerobic system.

2. The **phosphagen energy system** includes ATP and phosphocreatine, which has more energy than ATP.
   a. Phosphocreatine can donate phosphate groups to AMP and ADP to form ATP.
   b. The phosphagen energy system is used for short bouts of maximal exercise and provides the most rapidly available source of ATP for use by muscles.
   c. ATP and phosphocreatine are stored within the contractile elements of muscles.

3. The **glycogen-lactic acid system** occurs under anaerobic conditions and is used as a rapid source of energy for muscle contraction.

4. The **aerobic energy system** is the source of energy for prolonged light work.

B. Oxygen consumption during exercise corresponds to the intensity of exercise.

1. During exercise, oxygen consumption increases during the first few minutes, then reaches a steady-state in which uptake corresponds to demand.
   a. Oxygen consumption is often standardized in units of body weight (O_2 mL/kg/min) and averages 44 to 51mL/

kg/min for untrained males.

b. During exercise, O_2 consumption for untrained females is 20% less than for untrained males.

2. **Maximal oxygen consumption (VO_2max)** occurs at the point where the maximum for oxygen transport and uptake in tissues is reached and a plateau is attained.

3. VO_2max represents a person's maximal aerobic power and can be affected by age, activity level, and training.
   a. VO_2max decreases with age.
   b. Training increases VO_2max by increasing maximal cardiac output and blood arterial-venous difference.

C. *Oxygen recovery of exercise is the result of a temporary disruption in homeostasis.*

1. **Recovery oxygen** is often defined as the amount of oxygen above the resting level that must be taken into the body after exercise.
   a. During light exercise, recovery oxygen remains constant after a steady-state is reached.
   b. During heavy exercise, recovery oxygen builds up until the exercise ends.

2. The repayment of the recovery oxygen occurs in two stages: the **fast component** and the **slow component**.
   a. In the fast component stage, oxygen, which is needed to replenish the oxygen stores of the body and the phosphagen system, is usually repaid within two to three minutes after exercise.
   b. In the slow component stage, lactic acid is removed by conversion to pyruvic acid and glucose.

D. *Several nutrients fuel the work of exercise (carbohydrate, fatty acids, amino acids).*

1. The type of nutrient (carbohydrate, fatty acids, amino acids) used during exercise depends on the intensity and duration of the exercise and the availability of the stored fuels.
   a. At low rates of exercise of short duration, lipids supply a large proportion of the required energy.
   b. With increasing intensity and duration of exercise, a larger proportion of energy yield comes from carbohydrates.
   c. The choice of fuel is limited to carbohydrates and fats.

2. Diet affects:
   a. The types of nutrients available and used during exhaustive exercise.
   b. Level of endurance.

III. Respiration During Exercise

A. *Pulmonary ventilation during exercise is proportional to 0_2 consumption.*

1. Increased oxygen demands are paralleled by increased pulmonary ventilation.

2. The respiratory system is not normally the limiting factor in the delivery of oxygen to muscles during maximal exercise.

B. *The diffusing capacity of the lung during exercise increases as a result of increased lung blood flow.*

1. Diffusing capacity provides an index of:
   a. The dimensions of the pulmonary capillary bed.
   b. The pulmonary membrane.
   c. Overall efficiency of the system in gas exchange.

2. The increased **diffusing capacity** of oxygen and carbon dioxide in the lung during exercise results from increased pulmonary flow.

C. *Control of pulmonary ventilation during exercise involves three primary homeostatic mechanisms.*

1. Control of pulmonary ventilation involves:
   a. Interactions between the central nervous system inspiratory and expiratory centers.
   b. Pulmonary stretch receptors.
   c. Peripheral and central chemoreceptors sensitive to blood gas pressure and H^+ concentration.

2. The mechanisms controlling ventilation at rest cannot explain the increased ventilation or **hyperpnea** during exercise.

3. The time course of ventilation changes before and during moderate steady-state exercise can be divided into three components:
   a. A rapid rise occurring just before exercise starts.

b. A slower rise.
c. Development of **steady-state ventilation**.

4. Neural and chemical stimuli control pulmonary ventilation, but the mechanisms are unknown.
   a. During exercise, chemical stimuli include increased arterial [CO_2], increased arterial [H^+], and decreased arterial [O_2].
   b. During exercise, neural stimuli include impulses from the central nervous system centers (motor cortex), peripheral proprioreceptors, and lung stretch receptors and increased sensitivity of respiratory centers to arterial P_{CO2} and increased body temperature.
   c. The fast component is believed to be associated with neural stimuli.
   d. The slow components are believed to be related to chemical stimuli that alter central nervous system respiratory center activity.

5. P_{CO2} alone, lactic acid or H^+ concentration, and hypoxemia do not appear to provide the primary stimulus for increased ventilation during exercise.

6. Neural stimuli acting alone cannot explain increased ventilation during exercise or hyperpnea.

D. *Circulation during exercise increases to deliver more nutrients to the working muscles and to remove metabolic by-products.*

1. The increase in blood flow is attributed to several factors.
   a. About 50% of the increase to muscles results from local vasodilation caused by direct effects of increased muscle metabolism.
   b. The increase is also attributed to a moderate increase in arterial blood pressure.

2. Cardiac output parallels oxygen consumption and work output during exercise.
   a. An untrained individual can increase cardiac output to about four times the resting level.
   b. A trained individual can increase cardiac output about six times, mainly as a result of **hypertrophy** of the heart.

3. Stroke volume and heart rate increase during exercise.
   a. A trained athlete has a greater stroke volume and lower heart rate at rest and during maximum exercise than an untrained individual.
   b. Stroke volume reaches a maximum when cardiac output approaches 15 L/min. Further increases in cardiac output result from increases in heart rate during strenuous exercise.
   c. The capacity of the cardiovascular system is normally a limiting factor in terms of maximal oxygen consumption and exercise performance.

4. During exercise, blood flow distribution is altered in the following ways:
   a. Brain blood flow is maintained.
   b. Splanchnic and renal blood flow is decreased.
   c. Blood flow to skeletal muscles, heart, and skin is increased.

E. *Body temperature increases as a well-regulated response to exercise.*

1. The mechanical efficiency of the human body is less than 25%, and more than 75% of metabolic energy of exercise is converted to heat.

2. Body temperature rises during exercise and levels off at a few degrees above normal regardless of atmosphere temperature.
   a. Highest temperatures occur in exercising muscles.
   b. Temperature depends on the level of metabolism relative to the individual's maximal O_2 consumption.
   c. Less strain is placed on thermoregulatory mechanisms if body temperature rises during exercise.

3. Nonregulated increases in body temperature exceeding 42°C can cause heat stroke.

Key Terms

aerobic energy system
anaerobic
diffusing capacity
dynamic exercise (positive-negative)
glycogen-lactic acid system
heat stroke
humoral stimuli
hyperpnea
hypertrophy
hypoxemia
lactic acid
maximal aerobic power
maximal oxygen consumption (VO_2max)
phosphagen energy system

phosphocreatine
pulmonary stretch receptor
pulmonary ventilation
recovery oxygen
static exercise

Testing Your Knowledge

I. Matching. Select the correct term from the Key Terms.

________ 1. energy system used for short bouts of maximal exercise

________ 2. exercise performed during isometric muscle contraction

________ 3. plateau of oxygen uptake during exercise

________ 4. amount of gas moving across pulmonary membrane per minute per mm Hg pressure difference between alveolar air and pulmonary blood

________ 5. type of exercise quantified in foot-pounds or watts

________ 6. energy system producing most ATP

________ 7. amount of oxygen required after exercise

________ 8. term for increased ventilation during exercise

II. Dichotomous. Circle the correct choice from the paired terms in parentheses.

1. (Static/Dynamic) exercise is performed by a person pushing against a wall.

2. The (phosphagen/glycogen-lactic acid/aerobic) system produces the most ATP molecules.

3. During light exercise, energy is initially obtained (aerobically/anaerobically).

4. A trained female athlete has a (higher/lower) VO_2max than a sedentary male.

5. Subject A is a 40-year-old male marathon runner. Subject B is a 45-year-old male power walker. After one hour of exercise, the recovery oxygen for A would be (greater/less) than that for B.

6. After three hours of exercise, muscle recovery occurs most rapidly for a person on a high (fat/protein/carbohydrate) diet.

7. Phase (I/II/III) of hyperpnea during exercise is characterized by steady-state ventilation.

8. During exercise, (hyperthermia/hypothermia) occurs.

9. Compared to pre-exercise level, during strenuous exercise activity arterial O_2 (increases/decreases/changes little).

10. In a trained endurance athlete, hypertrophy is characterized by (normal/large) ventricular cavity with (normal/thick) ventricular walls.

III. True / False. To the left of the number write T for true or F for false.

_______ 1. Phosphocreatine may donate two phosphates to form two molecules of ATP.

_______ 2. Oxygen consumption increases linearly over time during the entire exercise period.

_______ 3. A sedentary 50-year-old man has a VO_2max comparable to an active 60-year-old man.

_______ 4. During the slow stage of the repayment of recovery oxygen, lactic acid is converted to pyruvic acid and glucose.

_______ 5. Pulmonary ventilation increases linearly with O_2 consumption in liters/minute.

_______ 6. Compared to baseline, arterial Pco_2 does not increase during moderate exercise.

_______ 7. During light exercise, significant increases in arterial [H+] occur.

_______ 8. The primary stimulus for exercise hyperpnea is increased stimulation of lung stretch receptors.

_______ 9. Increased arterial blood pressure is

largely responsible for increased muscle blood flow during strenuous exercise.

_______ 10. A well-trained athlete can increase cardiac output to about 30 L/min.

IV. Sentence Completion. Write your response in the space provided.

1. Climbing stairs is a type of ____________ exercise.

2. The standardized unit for oxygen consumption is ____________ .

3. Phosphocreatine is regenerated during the ____________ component stage of oxygen recovery.

4. A person who consumes a high ________ diet has greater endurance than a similar size person consuming a ___________ diet.

5. Muscle blood flow may increase ________ during maximal exercise as a result of local ____________ .

6. At rest and during maximal exercise, a trained athlete has a ____________ stroke volume and ____________ heart rate than the untrained person.

7. In a trained athlete during exercise, blood flow to the _________ and _________ increases about 4 times and 18 times, respectively, blood flow to the __________ remains constant, and blood flow to the ________ and ________ decreases.

8. The capacity of the __________ is normally the limiting factor in terms of maximal O_2 consumption and exercise performance.

9. Temperature increase in an exercising muscle is directly related to ___________ .

10. In an endurance-trained athlete at a cardiac output of 15 L/min, ________ is the major contributing factor, whereas at a cardiac output of greater than 25 L/min, ________ is the major contributing factor.

V. Multiple Choice. Circle the letter of the best response.

1. You are running a 50-yard dash. The energy for your muscles would come from the:
   a. phosphagen system
   b. glycogen-lactic acid system
   c. aerobic energy system
   d. glycogenolysis
   e. glycolysis

2. Your muscle has an oxygen deficiency. Which system for generating ATP would be most affected?
   a. phosphagen
   b. glycogen-lactic acid
   c. aerobic
   d. glycolysis
   e. gluconeogenesis

3. Which person would have the highest maximal aerobic power?
   a. a 30-year-old female marathon runner
   b. a 25-year-old male marathon runner
   c. a 25-year-old female body builder
   d. a 30-year-old male weight lifter
   e. a 40-year-old male triathlete

4. You are participating in a 10-km run to raise money for cancer research. After three hours, the source of energy for your leg muscles is:
   a. lipids
   b. glycogen
   c. glucose
   d. proteins
   e. all of the above

5. The change in pulmonary ventilation that occurs during exercise is due largely to an:
   a. increase in tidal volume
   b. increase in respiratory rate
   c. increase in VO_2max
   d. increase in diffusing capacity
   e. all of the above

6. The fast component of hyperpnea is triggered by:
   a. neural stimuli from motor cortex
   b. increase in arterial Pco2
   c. increase in arterial [H+]
   d. arterial hypoxemia
   e. stimulation of peripheral proprioceptors

7. The primary stimulus for exercise hyperpnea is:

a. arterial hypoxemia
b. increased arterial [H^+]
c. increased arterial P_{CO2}
d. signals from peripheral proprioreceptors
e. none of the above

8. During light to moderate exercise, all of the following parameters show little change from baseline (pre-exercise) except:
a. arterial P_{CO2}
b. arterial P_{O2}
c. diffusion capacity of the lung
d. arterial [lactic acid]
e. arterial [H^+]

9. Compared to a nonathlete at rest and during maximal exercise, which of the following parameters would be greater in a trained athlete under conditions of resting and maximal exercise?
a. stroke volume
b. cardiac output
c. heart rate
d. heart rate and stroke volume
e. stroke volume and cardiac output

10. In a comparison of blood flow distribution during rest and exercise in an athlete, the largest change in terms of blood flow distribution occurs to the:
a. digestive system
b. renal system
c. skeletal muscle
d. skin
e. heart

VI. Complete the Table.

For each condition, indicate the effect on oxygen uptake or consumption.

Condition	Effect on O_2 Uptake
1. Increased blood lactic acid	
2. Sedentary aging adult	
3. Six weeks of athletic training	
4. Decreased cardiac output	
5. Increased body temperature	

Chapter 31

Reproductive Physiology

Chapter Learning Objectives

1. Describe the male reproductive structures.
2. Describe the processes of spermatogenesis and spermiation and the role of the Sertoli cells.
3. Trace the pathway of sperm from its site of production to its exit from the body, and indicate the sources of fluids comprising semen.
4. Indicate the effects of follicle-stimulating hormone, luteinizing hormone, and testosterone in the male; describe the relationship and feedback loop between the hypothalamus, anterior pituitary, and testes regarding the production of sperm and testosterone.
5. Describe the female reproductive system.
6. Diagram the process of oogenesis.
7. Describe the events of the follicular, luteal, and menstrual phase of the ovarian cycle.
8. Explain the action of gonadotropins on the female reproductive system; describe the relationship and feedback loop of the hypothalamus, anterior pituitary and ovaries in the production of hormones, mature ova, and ovulation.
9. Indicate the roles of estrogen and progesterone maturation of the follicle and on the uterine endometrium.
10. Explain what happens during menopause.

Expanded Chapter Outline

I. Reproductive Physiology in the Male

A. *The reproductive anatomical structures in the male include both the internal and the external genitalia.*

1. The external reproductive organs are the **penis** and **scrotum**, which holds the **testes**, **epididymis**, and spermatic cord.
   a. Sperm reproduction requires a temperature lower than that of the body.
   b. Cremaster muscles in the scrotum respond reflexively to temperature changes.

2. The testes have two major functions:
   a. Production of sperm.
   b. Production of male sex hormones.

3. Each testis is composed of two areas:
   a. The **seminiferous tubules** in which sperm are produced.
   b. The interstitial space that contains blood vessels, lymphatics, nerve fibers, connective tissue, extracellular fluid, and the testosterone-producing **Leydig cells**.

B. ***Spermatogenesis**, the formation of sperm, takes place in the seminiferous tubules of the testes.*

1. Spermatogenesis begins at puberty and continues throughout the male's life. The process takes 70 to 80 days.

2. Two types of cells are involved in spermatogenesis:
   a. Spermatogenic cells from which mature sperm arise.
   b. **Sertoli cells**, which provide mechanical and nutritional support for the spermatogenic cells.

3. **Spermatogonia** cells either divide mitotically to replenish their supply or differentiate into primary spermatocytes, which then divide meiotically and give rise to sperm.
   a. Spermatogonial cells are located along the basement membrane of the seminiferous tubules.
   b. During **meiosis**, homologous chromosomes exchange chromosomal segments in a process known as crossing over.

4. Spermatogenesis involves the following steps:

spermatogonium (46 chromosomes) —enlarges→ primary spermatocytes (46 chromosomes) primary spermatocyte —(1st meiotic division)→ two secondary spermatocytes

each secondary spermatocyte (23 chromosomes) —(2nd meiotic division)→ two spermatids (23 chromosomes)

each spermatid (23 chromosomes) —(spermiation)→ one sperm (23 chromosomes)

5. As a result of a complex series of cytological changes, spermatids differentiate into mature sperm, which enter the lumen of the seminiferous tubule.

6. Each sperm consists of:
   a. A head containing chromatin in the nucleus and an acrosome containing digestive enzymes.
   b. The tail or flagellum.
   c. The body or midpiece, which contains mitochondria.

C. *Sertoli cells serve several functions that are important for spermatogenesis.*

1. They are joined to each other by tight junctions, which form a blood-testes barrier.

2. They secrete fluid containing proteins, nutrients, androgens, and ions into the lumen to nourish sperm.

3. Sertoli cells synthesize two proteins:
   a. **Androgen binding protein (ABP),** which binds to **testosterone** and increases its concentration.
   b. **Inhibin,** which regulates the secretion of follicle-stimulating hormone.

D. *Sperm are transported through the internal reproductive structures.*

1. The pathway for sperm movement is:

lumen of seminiferous tubule → rete testis → epididymis → vas deferens → ejaculatory duct → urethra

   a. The **epididymis** secretes a fluid containing nutrients, enzymes, and hormones to aid sperm maturation and motility.
   b. Sperm transit through the epididymis takes about two weeks.
   c. The wall of the **vas deferens** is composed of smooth muscle and innervated by sympathetic nerves.
   d. The ejaculatory duct runs through the prostate gland.

2. **Seminal vesicles** secrete a viscous material rich in prostaglandins, fructose, and nutrients.
   a. Secretory ducts from the seminal vesicles empty into each vas deferens, where it joins the ejaculatory duct.
   b. Seminal fluid makes up about 60% of the total volume of ejaculated semen.

3. The **prostate gland** secretes an alkaline liquid, containing citric acid, calcium, and various enzymes.
   a. The fluid neutralizes the acidity in the vagina and increases sperm motility.
   b. Valves in the prostate prevent mixing of urine and semen.

4. **Semen** containing sperm and the secretions from the seminal vesicles and prostate passes from the ejaculatory duct into the urethra, where it is joined by secretions from the bulbourethral glands.

5. Ejaculation results from stimulation by sympathetic nerves of the smooth muscle in the epididymis, vas deferens, prostate, and seminal vesicles.

E. *Neuroendocrine control of reproduction in the male involves an interaction between the hypothalamus, pituitary, and testes.*

1. **Follicle-stimulating hormone (FSH)** stimulates the conversion of spermatogonia into primary spermatocytes.

2. **Luteinizing hormone (LH)** stimulates Leydig cells to synthesize and secrete testosterone.

3. The production of FSH and LH occurs via the sequence:
   a. Hypothalamus secretes **gonadotropin-releasing hormone (GnRH).**
   b. GnRH stimulates the anterior pituitary to produce and release FSH and LH.

4. The constancy of secretions of GnRH, LH,

and FSH is due to the presence of a continuous feedback loop between the hypothalamus, anterior pituitary, and testes.

5. Sensory stimuli (visual, olfactory, tactile) can alter the release of GnRH from the hypothalamus and hence the release of LH, FSH, and testosterone.

F. ***Testosterone**, the major sex hormone in the male, has a biological effect on many tissues of the body.*

1. Testosterone produces local effects on sperm production in the seminiferous tubules.

2. In the blood, testosterone is bound to albumin and to another plasma protein, **sex steroid-binding globulin (SSBG)**.

3. Testosterone acts directly on some tissues (muscle, kidney, and bone).

4. Testosterone is converted into another steroid before acting on tissues.
   a. It is transformed into dihydrotestosterone before acting on tissues such as the prostate and seminal vesicles.
   b. In the brain, testosterone is converted into estrogen.

5. At puberty, the hypothalamus secretes increasing amounts of GnRH.
   a. The increase in testosterone results in the enlargement of the penis, scrotum, and testes and stimulates the development of male secondary sex characteristics.
   b. Anatomical changes stimulated by testosterone mark the onset of puberty.

6. Spermatogenesis requires the presence of high levels of testosterone in the seminiferous tubules.
   a. Testosterone stimulates the formation of spermatogonia and spermatids.
   b. It stimulates protein synthesis and fluid secretion in Sertoli cells, which also secrete androgen-binding protein (ABP) to bind the hormone.

II. Reproductive Physiology in the Female

A. *The reproductive anatomical structures in the female are internal.*

1. The external genitalia, collectively called the **vulva**, include the labia minora, labia majora, the clitoris, and the vaginal opening.
   a. The labia minora is highly vascularized, and the glans clitoris contains a dense supply of nerve endings.
   b. The penis and **clitoris** have the same embryological origin.

2. The internal reproductive organs include the vagina, uterus, fallopian tubes, and ovaries.
   a. The **ovaries** produce ova, the germ cells, and hormones and have an embryonic origin similar to the testes.
   b. The uterine endometrium varies in thickness with the menstrual cycle.

B. ***Oogenesis**, the formation of ova, takes place in the ovaries.*

1. Meiosis begins in some oogonia during the second and third months of embryonic/fetal development.

2. The sequence of events in oogenesis is:
   a. Oogonia develop into primary **oocytes**; each is surrounded by a layer of follicle cells and called a primary follicle.
   b. **Zona pellucida** forms between the maturing ovum and stratum granulosum.
   c. Follicle cells increase in size and number to form the **stratum granulosum**, which provides nutrients, signals to promote further maturation, and forms a barrier against substances that could destroy the developing follicle.
   d. The highly vascularized **theca interna** develops from connective tissue surrounding follicle during the fifth to sixth months of fetal development.
   e. Further follicular development ceases until ovulation. Primary oocyte is arrested in first meiotic division.

3. Major differences between growth patterns of germ cells in the male and female are:
   a. The length of time for maturation of the sperm or ova from the primary spermatocyte or oocyte.
   b. Onset of the process.
   c. Number of mature gametes produced in a lifetime.
   d. Number of gametes released at one time.

4. Of the 7 million oocytes formed in the ovary during embryonic development, only

about 400 will reach maturity and be released from the ovary. The rest degenerate.

C. ***Ovulation**, the release of an ovum from one of two ovaries, occurs once each month from puberty until menopause.*

1. Growth of the follicle occurs in a series of steps.
   a. Each month 6 to 20 primary follicles begin to develop, but only one matures and becomes a Graafian follicle.
   b. A mature follicle is surrounded by granulosa cells, which secrete fluid containing proteins, electrolytes, polysaccharides, and hormones.
   c. The **antrum** forms around the ovum and fills with fluid.
   d. The mature follicle forms a bulge on the outer surface of the ovary.
   e. Around ovulation the follicle secretes proteolytic enzymes, which break down the ovarian surface.
   f. The **ovum** released at ovulation into the peritoneal cavity is drawn into the fallopian tube by ciliary action and fimbrial movement.
   g. Just before ovulation, meiosis resumes and the nucleus is arrested at the second meiotic division.
   h. Fertilization is the trigger for the ovum to complete the second meiotic division.

2. Growth of the corpus luteum occurs after ovulation.
   a. The **corpus luteum** forms from the remaining follicle cells (granulosa and thecal cells) after ovulation and secretes estrogen and progesterone.
   b. The corpus is highly vascularized for 7 to 8 days postovulation.
   c. If fertilization and implantation do not occur, the corpus luteum degenerates between days 8 to 21 following ovulation.

D. *Neuroendocrine control of reproduction in the female involves an interaction between the hypothalamus, pituitary, ovaries, and uterus.*

1. A major difference between hormonal control of reproduction in the male and female is the cyclical secretion of hormones and nature of the reproductive process in the female.

2. The hypothalamus secretes GnRH, which stimulates the anterior pituitary to produce FSH and LH that stimulate germ cell development and hormone secretion in the ovaries.

3. The effects of LH and FSH vary during different **phases of the ovarian cycle**. The three phases are the:
   a. **Follicular** (lasting about 12 days), characterized by maturation of the ovum.
   b. **Luteal** (lasting about 11 days), characterized by development of the corpus luteum.
   c. **Menstrual** (lasting about 5 days).

4. The **follicular (preovulatory) phase** occurs just after menstruation and is stimulated by FSH.
   a. FSH stimulate hypertrophy and hyperplasia of granulosa cells.
   b. Granulosa cells (in turn) secrete estrogen, which stimulates GnRH secretion and acts on the granulosa cells to increase the number of receptors for both estrogen and FSH.
   c. Estrogen has an autocrine action on granular cells and the developing follicle.
   d. FSH and estrogen cause an increase in the number of LH receptors in granulosa cells.
   e. Estrogen stimulates the pituitary to release LH, which stimulates granulosa cells to secrete progesterone, and the theca cells to synthesize and secrete testosterone and androstenedione, which exert a paracrine effect on cells in the follicle.
   f. Follicles not completing maturation undergo **atresia**.

5. The preovulatory surge of LH induces changes in follicular structure, resulting in ovulation.
   a. Peak levels of LH and FSH, referred to as preovulatory surges, occur just before ovulation.
   b. LH-induced changes include increase in production of antral fluid, protrusion of the follicle against the wall of the ovary, enzymatic degradation of the ovarian wall, and release of antral fluid and the ovum (ovulation).
   c. **Ovulation** marks the end of the follicular phase and occurs about day 14 of a 28-day cycle.

6. In the **luteal (postovulatory) phase,** the corpus luteum forms from granulosa cells in a process called luteinization that is stimulated by LH.
   a. Estrogen and progesterone secreted during the luteal phase inhibit GnRH, LH, and FSH secretion leading to falling levels of FSH and LH and inhibition of maturation of oocytes.
   b. Degeneration of the corpus luteum results in falling levels of estrogen and progesterone that releases the inhibition of GnRH, LH, and FSH.
   c. As blood levels of these hormones rise, a new follicular cycle begins.
   d. The mechanism responsible for degeneration of the corpus luteum is not known and may involve prostaglandins secreted by the corpus.
   e. The corpus luteum begins to degenerate on day 8 postovulation unless fertilization occurs.

7. In the **menstrual phase**, estrogen stimulates proliferation of the uterine endometrium and the number of progesterone receptors in the uterus.
   a. Estrogen and progesterone secreted in the luteal phase stimulate growth and proliferation of endometrial blood vessels and the accumulation of glycogen and enzymes in endometrial glands.
   b. In the absence of estrogen and progesterone, hemorrhagic changes occur in the endometrium, which becomes necrotic, and menstrual flow occurs.
   c. Prostaglandins produced by the uterus stimulate uterine muscle contractions to discharge blood.
   d. Menstrual phase lasts about 5 to 7 days.

E. *Estrogen, the primary sex hormone in the female, has a biological effect on many tissues of the body.*

1. **Estrogen** stimulates the development of sex organs and the appearance of secondary sex characteristics.

2. Estrogen interacts with a variety of other hormones that control metabolic processes. The effects of estrogen include:
   a. Decreased sensitivity of peripheral tissues to insulin.
   b. Increased levels of plasma proteins binding thyroxine, cortisol, and progesterone.
   c. Increased circulating levels of renin.
   d. Protection against bone loss.

3. **Progesterone** is an androgen antagonist and prevents target tissues from responding to normally low levels of circulatory androgens in the female by competing for androgen-binding sites in androgen-dependent tissues.

F. ***Menopause*** *refers to the cessation of menstruation in the female.*

1. It occurs on the average at age 50, but women exhibit much variation in onset.

2. Postmenopausal estrogen derives from estrogen precursors from the adrenal cortex.

3. Menopause is characterized by loss of vaginal epithelium, decrease in breast mass, increase in bone loss, hot flashes, and increased risk of cardiovascular disease.

Key Terms

acrosome
amenorrhea
antrum
atresia
blood-testes barrier
clitoris
corpus luteum
dysmenorrhea
endometrium
epididymis
estrogen
follicle-stimulating hormone
follicular phase
gonadotropin-releasing hormone
granulosa cells
Leydig cells
luteal phase
luteinizing hormone
menopause
menstrual phase
oogenesis
ovarian cycle
ovulation
paracrine
primary follicle
progesterone
prostate
seminiferous tubules
Sertoli cells
spermatogenesis
spermiation
testosterone

Testing Your Knowledge

I. Matching. Select the correct term from the Key Terms.

________ 1. releases testosterone

________ 2. secretes progesterone

________ 3. secretes estrogen

________ 4. site of sperm production

________ 5. triggers ovulation

________ 6. stimulates sperm production and maturation of primary follicle

________ 7. produces alkaline fluid

________ 8. part of vulva homologous to penis

________ 9. absence of menstruation owing to pregnancy

________ 10. formed by tight junctions between Sertoli cells

________ 11. stimulates anterior pituitary to release FSH and LH

________ 12. fluid-filled space in Graafian follicle

________ 13. uterine layer shed during menstrual phase

________ 14. coiled tube in which sperm mature

________ 15. part of sperm containing digestive enzymes

II. Dichotomous. Circle the correct choice from the pair terms in parentheses.

1. In a cross section through a seminiferous tubule, primary spermatocytes would be found near the (periphery/lumen) of the tubule.
2. The transformation of spermatids to sperm is called (spermatogenesis/spermiogenesis).
3. (Smooth/Skeletal) muscle is found in the epididymis, scrotum, and vas deferens and is innervated by the (somatic/sympathetic) nervous system.
4. (FSH/LH) is necessary for production of testosterone by (Sertoli/Leydig) cells.
5. (One ovum/Four ova) is (are) produced from an oogonial cell.
6. During the (follicular/luteal) phase, estrogen stimulates release of GnRH and FSH.
7. Estrogen has a(n) (autocrine/paracrine) action on granulosa cells, and testosterone has a(n) (autocrine/paracrine) action on Sertoli cells.
8. The effect of testosterone on the production of gonadotropins is comparable to the effect of estrogen on the production of gonadotropins that occurs in the (follicular/luteal phase).
9. Leydig cells in the male are comparable to (theca/granulosa) cells in the female as to the effect of LH.
10. Sertoli cells are functionally similar to the (granulosa/oogonial) cells in females.

III. True / False. To the left of the number write T for true or F for false.

________ 1. Sperm may be destroyed by antibodies.

________ 2. Sperm removed from the epididymis are capable of fertilizing an ovum.

________ 3. Testosterone is transported in blood by androgen-binding protein.

________ 4. All primary oocytes eventually become mature ova.

________ 5. At the time the corpus luteum is active, the level of LH would be at its highest.

________ 6. The first meiotic division of oogenesis is completed during the follicular phase of the ovarian cycle.

________ 7. Estrogen is synthesized and released by both the granulosa and the theca cells of the follicle.

________ 8. The endometrium becomes thicker and more vascularized as a result of the direct action of FSH and LH.

________ 9. In both males and females, FSH and LH act directly on spermatogonia and oogonia cells to complete their maturation.

________ 10. A postmenopausal woman produces no sex hormones.

IV. Sentence Completion. Write your response in the space provided.

1. One primary spermatocyte gives rise to __________ (number) spermatid(s) and one primary oocyte gives rise to __________ (number) ovum/ova.

2. If a spermatid has 20 chromosomes, a spermatogonial cell in the same organism has __________ chromosomes.

3. The blood level of testosterone is ultimately controlled by the level of ____________ secreted by the __________.

4. In a 40-year-old woman, the primary follicles are __________ years old.

5. LH stimulates granulosa cells to secrete __________ and theca cells to secrete ____________.

6. In a 28-day ovarian cycle, ovulation would occur on day ____________ .

7. During the luteal phase, estrogen and progesterone ______________ secretion of GnRH, FSH, and LH.

8. During the luteal phase, estrogen and progesterone are synthesized and secreted from the ________________, which develops from the ______________ cells.

9. You are viewing a cross section of a human female's ovary and see three corpora lutea. This female released __________ at her last ovulation.

10. The penis is embryologically homologous to the __________ in females.

V. Multiple Choice. Circle the letter of the best response.

1. The pathway for sperm from the testes to outside the body is:
   a. rete testis → vas deferens → ejaculatory duct → ureter
   b. epididymis → spermatic cord → vas deferens → ejaculatory duct
   c. epididymis → ejaculatory duct → vas deferens → urethra
   d. vas deferens → epididymis → ejaculatory duct → urethra
   e. epididymis → vas deferens → ejaculatory duct → urethra

2. The cells in your reproductive system that are genetically identical to your body cells are:
   a. spermatogonia
   b. secondary spermatocytes
   c. spermatids
   d. primary spermatocytes preparing for cell division
   e. all of the above

3. Which set of paired terms is a mismatch?
   a. spermatid sperm: spermatogenesis
   b. primary oocyte ovum: oogenesis
   c. Golgi apparatus: acrosome
   d. mitochondria: midpiece
   e. tight junctions: blood-testes barrier

4. Testosterone levels are maintained at a high concentration in the seminiferous tubules because of the:
   a. FSH
   b. LH
   c. inhibin
   d. ABP
   e. SSBE

5. In a 25-year-old woman with a 28-day cycle, the levels of estrogen and progesterone would be lowest on days:
   a. 1-6
   b. 7-13
   c. 14-20
   d. 21-28

6. The level of progesterone is highest during (the):
   a. luteal phase
   b. follicular phase
   c. menstrual phase
   d. day of ovulation

7. A woman with impaired function of the anterior pituitary would exhibit all of the following symptoms except:
   a. decreased urine output
   b. menstrual irregularities
   c. decreased metabolic rate
   d. infertility
   e. none of the above

8. In a woman taking oral contraceptives:
   a. the level of FSH increases during the follicular phase
   b. the level of LH increases during the follicular phase
   c. the anterior pituitary does not release its gonadotropic hormones
   d. the corpus luteum does not degenerate
   e. more androgens are produced

9. Under which conditions would testosterone in the male and estrogen in the female increase?
   a. high level of progesterone and testosterone
   b. low level of FSH, high level of LH
   c. low level of LH, high level of FSH
   d. high level of LH
   e. low level of estrogen and testosterone

10. In males, in the absence of hypothalamic gonadotropic-releasing hormones:
   a. neither sperm nor testosterone is produced
   b. the anterior pituitary produces FSH but not LH
   c. the testes produce testosterone but no sperm
   d. the adrenal cortex increases its production of estrogen
   e. the anterior pituitary produces larger quantities of FSH and LH

VI. Complete the Table.

For each condition indicate the effect (increase, decrease, none) on oogenesis and/or spermatogenesis.

Condition	Effect on Oogenesis	Effect on Spermatogenesis
1. Decrease in testosterone		
2. Absence of fertilization		
3. Severing of hypothalamic-pituitary portal		
4. Absence of Leydig cells		
5. Absence of spermatogonia		
6. Continuous presence of estrogen or progesterone (or both)		
7. Increase in LH		
8. Absence of androgen-binding protein		
9. Excess of inhibin		
10. Removal of prostate		

VII. Complete the Table.

For each condition, indicate the effect (increase, decrease, none) on the production of the hormones.

Condition	GnRH	FSH	LH	Testosterone	Estrogen
1. Excess inhibin					
2. Excess testosterone					
3. Decrease in GnRH					
4. Increase in FSH					
5. Increase in LH					
6. Increase in granulosa cells					
7. Degeneration of corpus luteum					
8. Presence of estrogen during follicular phase					
9. Presence of estrogen during luteal phase					
10. Presence of progesterone					

Chapter 32

Pregnancy, Fetal Development, and Lactation

Chapter Learning Objectives

1. Give the details of fertilization; describe the barriers surrounding the ovum and sperm penetration.
2. Describe two methods by which polyspermy is prevented.
3. Explain the phenomenon of multiple births.
4. Describe implantation and early cleavage.
5. Describe the formation, structure, and function of the placenta; describe the hormones produced by the placenta and their effects on the mother and fetus.
6. Describe the physical and physiological changes that occur in the pregnant woman during each trimester.
7. Describe the growth and development of the fetus.
8. Tell what happens at birth; describe the effects of oxytocin, prostaglandins, progesterone, and relaxin on parturition.
9. Describe the anatomy of the breast; the roles of estrogen, progesterone, prolactin, and oxytocin; and the effects of suckling on milk production and release.

Expanded Chapter Outline

I. The Process of Fertilization

A. *Following ovulation, the ovum is transported through the fallopian tube to the uterus.*

1. The ovum is drawn into the fallopian tube by the sweeping motion of cilia and fimbrial movements.

2. Fertilization normally occurs in the fallopian tube.

B. *Following ejaculation into the vagina, sperm are transported by way of the vagina, cervix, and uterus into the fallopian tubes.*

1. Sperm are propelled to the fallopian tubes by wavelength movements of their tails and by contractions of muscles in the uterus and fallopian tubes.

2. The volume and consistency of cervical mucus changes during the ovarian cycle.
   a. At ovulation cervical mucus is clear and watery.
   b. Glycoproteins in the mucus form elongated fibers to allow sperm to penetrate the mucus and pass upward through the cervix.
   c. Before and after ovulation, cervical mucus is thick and sticky and lacks glycoprotein channels.

3. An ejaculate contains an average of 100 to 500 million sperm, but up to 50% may be incapable of fertilization because of abnormal shape or reduced motility.
   a. One is required for fertilization.
   b. Sperm are destroyed by acidic vaginal secretions.

C. *Sperm must undergo the acrosomal reaction before they are capable of fertilizing an ovum.*

1. In the epididymis, sperm mature and acquire the ability to move but not to fertilize the ovum.

2. Sperm must undergo **capacitation** in the female reproductive tract before fusion with the ovum.

3. The mature ovum is surrounded by the zona pellucida and several layers of granulosa cells.

4. The **acrosomal reaction** prepares the sperm for penetration of the zona pellucida surrounding the ovum through the release of proteolytic (digestive) enzymes to digest the zona pellucida.

D. *Fertilization refers to the fusion of sperm and ovum.*

1. When the sperm reaches the ovum membrane, it is engulfed by the ovum.

2. **Polyspermy** is prevented electrically and chemically.
   a. Initially, it is prevented by an electrical change in the ovum's surface immediately after one sperm has been engulfed.
   b. Later, a chemical change occurs involving the release of cortical granules into the space between the plasma membrane and zona pellucida.

E. *Multiple births may be monozygotic, dizygotic, or a combination of both ovum.*

1. Fertilization of two or more ova released at ovulation results in genetically different dizygous twins, triplets, or quadruplets.

2. Monozygous twins are genetically identical (triplets or quadruplets) and result from the fertilization of a single ovum by a single sperm followed by a special type of cell division.

II. Cell Division and Implantation Following Fertilization

A. *The fertilized ovum undergoes cell division during transport through the fallopian tube.*

1. Female and male **pronuclei** join to form the diploid nucleus of the zygote.

2. The **zygote**, the fertilized ovum containing 46 chromosomes, begins to divide mitotically about 30 hours postfertilization.
   a. The eight-cell stage is called a morula.
   b. As cell division continues, a hollow area develops in the cell mass, and the morula becomes a **blastocyst**.
   c. The cell divisions leading to the blastocyst occur while the cell mass is being moved down the fallopian tube toward the uterus.

4. The blastocyst is composed of two types of cells:
   a. **Trophoblast cells**, which play a role in **implantation** and later give rise to the placenta.
   b. **Embryoblasts**, which give rise to the embryo proper.

B. *The blastocyst implants in the endometrium of the uterus.*

1. The blastocyst secretes estrogen and proteolytic enzymes, which digest the uterine endometrium to allow implantation.
   a. Endometrial cavities formed by the **decidual reaction** fill with blood.
   b. The trophoblast cells secrete **human chorionic gonadotropin (hCG)**, which prevents degeneration of the corpus luteum.
   c. Estrogen secreted by the corpus luteum stimulates the secretion of prostaglandins from the endometrium.
   d. Progesterone stimulates uterine endometrium to receive and nourish the blastocyst.

2. The trophoblast cells penetrate the endometrium and differentiate into two layers: **amnion** and **chorion**.
   a. The fluid-filled amniotic cavity protects the embryo.
   b. The chorion gives rise to the fetal part of the **placenta**.

3. Implantation is completed around day 9 following ovulation.

4. Ectopic pregnancies may result from abnormal ovum transport in the fallopian tube and abnormal implantation.

III. The Placenta

A. *The embryo is dependent on the maternal circulation for delivery of nutrients and oxygen and for removal of waste products.*

B. *The placenta serves as an interface between the fetal and maternal circulatory systems.*

1. The placenta is formed from trophoblast cells and the uterine endometrium.
   a. Villi from the developing embryo invade pools of maternal blood in the endometrium.
   b. Capillaries in the villi serve as the route for exchange between mother and embryo.
   c. Small blood vessels become the arteries and veins of the umbilical cord and placenta.
   d. No mixing of fetal and maternal blood occurs.
   e. The fetal flow of blood through the placenta is:

umbilical artery → perfusion through placental capillaries → **umbilical vein**

2. Exchange of gases (O_2 and CO_2) occurs in the placenta.
   a. Fetal blood carries more oxygen than maternal blood because the concentration of hemoglobin is 50% higher in the fetus, and fetal hemoglobin can carry 20% to 30% more oxygen than maternal hemoglobin.
   b. CO_2 is exchanged by simple diffusion.

3. Some nutrients such as glucose, potassium, and sodium diffuse across placental membranes, whereas others such as proteins are extracted.

4. Fetal wastes such as urea, uric acid, and nonprotein nitrogens pass by simple diffusion from fetal to maternal blood in the placenta.

5. Drugs and toxic chemicals may reach the fetus via the placenta and affect its growth and development.

6. The placenta actively and selectively transports large maternal antibodies to the fetus.

C. The placenta and embryo are protected from immunological rejection.

1. Rejection does not occur because:
   a. The blastocyst does not express antigens.
   b. The placenta secretes human chorionic gonadotropin, which suppresses lymphocytes.
   c. The embryo is protected by membranes.

2. Maternal and fetal blood may mix during parturition, which may lead to Rh incompatibility problems in subsequent pregnancies.

IV. Hormonal and Physical Changes in the Female During Pregnancy

A. The placenta secretes four hormones that are important for the continuation of the pregnancy.

1. Placental hormones are:
   a. **Human chorionic gonadotropin (hCG).**
   b. Estrogen and progesterone.
   c. **Human chorionic somatomammotropin (hCS),** also called human placental lactogen (hPL).

2. hCG, produced first by trophoblast cells and later by the placenta, stimulates the corpus luteum to continue secreting estrogen and progesterone to maintain the uterine endometrium.

3. During the second month, the placenta stops producing hCG and secretes estrogen and progesterone so that the corpus luteum is no longer necessary.
   a. The placenta secretes ten times more progesterone than the corpus luteum.
   b. The placenta extracts cholesterol from maternal circulation to produce progesterone.

4. Progesterone is essential for maintenance of pregnancy.
   a. It stimulates growth and secretion of nutrients from the endometrium.
   b. It prepares the breast for lactation.
   c. It inhibits uterine contractions.
   d. It serves as a precursor for estrogen synthesis by the **fetoplacental unit.**

5. Estrogen stimulates enlargement of the uterus and external genitalia, breast growth, and relaxation of pelvic ligaments.

6. Estrogen synthesis requires a complex interaction between the placenta and fetal adrenal gland.
   a. The fetal adrenal glands contain enzymes needed to convert cholesterol to androgens, specifically **dehydroepiandrosterone-sulfate (DHEA-sulfate).**
   b. DHEA-sulfate is transported to the placenta, where it is converted to estrogen.

7. **hCS** plays a role in metabolism and nutrition of both the pregnant woman and her fetus.
   a. hCS is chemically similar to human growth hormone.
   b. It decreases maternal utilization of glucose and promotes the release of free fatty acids from maternal stores.

B. In addition to the hormones secreted by the feto placental unit, a number of hormonal changes occur in the pregnant female.

1. Secretion of hypothalamic GnRH and pituitary gonadotropins (LH and FSH) is inhibited during pregnancy.

2. The pituitary secretes **prolactin,** a hormone

essential for synthesis of breast milk.

3. **Relaxin** is secreted initially by the corpus luteum and later by the uterine endometrium.
   a. It relaxes pelvic ligaments.
   b. With progesterone, relaxin inhibits uterine contractions.

4. Insulin secretion increases after the third month in response to decreased maternal sensitivity to the hormone.

5. An increase in secretion of aldosterone maintains adequate sodium and water retention during pregnancy to maintain high maternal and fetal blood volumes.

6. Levels of free and protein-bound cortisol increase as a result of increased secretion by maternal adrenal glands and an increase in cortisol binding globulin (CBG) in maternal plasma.
   a. CBG synthesis by the liver increases in response to estrogen stimulation.
   b. Increased cortisol contributes to an increase in maternal adipose tissue and to development of the mammary glands.

7. Increased maternal thyroid function stimulates production of thyroxine (T_4) and triiodothyronine (T_3) during pregnancy.
   a. T_3 and T_4 stimulate maternal cardiopulmonary function.
   b. The rate of delivery of oxygen and nutrients to the fetus is increased.

C. Pregnancy is divided into three-month periods called trimesters, each marked by dramatic changes in many tissues of the body.

1. Physical and physiological changes during the **first trimester** may include morning sickness (tiredness, nausea, vomiting), frequent urination, breast swelling and tenderness, and increased vaginal secretions.

2. Physical and physiological changes during the **second trimester** may include fluid retention that results in edema of hands and feet, increased production of red blood cells, increased appetite and maternal cardiac output, and digestive system changes.

3. Physical and physiological changes during the **third trimester** may include large weight gain, frequent urination, muscle cramps, uterine contraction, disturbed sleep patterns, increased fatigue, and shortness of breath.

D. Development of the fetus is divided into three stages: the germinal, embryonic, and fetal.

1. The **germinal stage** is from fertilization to two weeks, the embryonic stage from two to eight weeks, and fetal stage from eight weeks to birth.

2. In the germinal stage, the trophoblast cells give rise to the inner amnion and outer chorion, and the inner cells of the blastocyst differentiate into three layers from which all body structures develop.
   a. The outer layer gives rise to skin, sense organs, and nervous systems.
   b. The middle layer gives rise to the skeleton, muscles, heart, and blood vessels.
   c. The inner layer gives rise to organs of the digestive system, liver, and pancreas.

3. During the **embryonic stage**, all of the major organs become partially formed: limb buds, eyes, ears, a beating heart, and organs of the digestive and nervous systems.

4. During the **fetal stage**, all rudimentary organs and structures become more differentiated and begin to function.
   a. In the eighth month, many organs mature, e.g., the lungs produce surfactant.
   b. In the ninth month, the fetus positions itself head down in preparation for parturition.

V. Parturition

A. ***Parturition****, or delivery, involves both cervical dilation and uterine contractions, which are under hormonal control.*

1. **Oxytocin** stimulates contraction of uterine muscles.
   a. **Cervical dilation** triggers release of oxytocin from the posterior pituitary.
   b. The number of oxytocin receptors in the uterus increases during the later part of pregnancy.

2. **Prostaglandins** stimulate contractions of uterine smooth muscle.
   a. Oxytocin stimulates release of prosta-

glandins from the myometrium.
   b. Prostaglandin exerts a paracrine action.

3. The level of progesterone that inhibits uterine contractions drops just before delivery.

4. Relaxin increases the number of oxytocin receptors in the uterus, softens the cervix, and causes relaxation of pelvic ligaments.

B. Lactation, the synthesis and release of breast milk, is under hormonal control.

1. The breast is composed of glandular epithelium and a ductal system embedded in interstitial tissue and fat.
   a. The ducts that terminate in the nipple are connected to alveoli.
   b. The **alveoli**, the milk-secreting glands, are surrounded by myoepithelial cells.

2. Breast development during pregnancy is stimulated by a number of hormones, including estrogen, progesterone, hCG, hCS, prolactin, and growth hormone.
   a. Breasts increase in size, and the ductal and alveolar structures become more complex.
   b. Alveolar cells extract nutrients from the maternal circulation and convert these into breast milk.

3. Estrogen and progesterone stimulate breast development and release of prolactin but inhibit milk production before parturition.

4. Prolactin and oxytocin control the production and release of milk following parturition.
   a. Prolactin levels begin to decline after birth.
   b. Suckling of the breast stimulates release of prolactin and oxytocin via neural pathways to the hypothalamus.
   c. Prolactin levels rise and fall with every episode of nursing.
   d. Prolactin inhibits the release of GnRH, LH, and FSH.
   e. Oxytocin stimulates the release of milk from the breast by acting on the myoepithelial cells around alveoli.

5. Breast milk contains proteins (largely lactalbumin), water, lactose, fat, and a variety of vitamins and minerals.

Key Terms

acrosome
acrosome reaction
alveoli
amnion morula
areola
blastocyst, morula
capacitation
cervical dilation
cervical mucus
chorion
colostrum
corpus luteum
cortical granule
cortisol binding globulin (CBG)
dizygotic twins
ectopic pregnancy
embryoblasts
embryonic stage
endometrium
fallopian tube
germinal stage
human chorionic gonadotropin (hCG)
human somatomammotropin
implantation
lactation
lactalbumin
monozygotic twins
morning sickness
morula
myoepithelial cells
oxytocin
placenta
polyspermy
prolactin
pronucleus
prostaglandin
relaxin
surfactant
trimester
trophoblast
umbilical cord
zona pellucida

Testing Your Knowledge

I. Matching. Select the correct term from the Key Terms.

________ 1. released by ovum as a block to polyspermy

________ 2. forms maternal part of placenta

________ 3. results from abnormal implantation site

________ 4. component of milk

________ 5. released from posterior pituitary and stimulates uterine contractions

________ 6. forms fluid-filled sac around fetus

________ 7. process by which sperm attain ability to fertilize ovum

________ 8. site of fertilization

________ 9. physical change in mother occurring during first trimester

_________ 10. part of sperm containing digestive enzymes

_________ 11. individuals with exactly the same DNA

_________ 12. give rise to embryo proper

_________ 13. give rise to fetal part of placenta

_________ 14. interface between maternal and fetal circulatory systems

_________ 15. secreted by blastocyst and stimulates corpus luteum to continue producing its hormones.

II. Dichotomous. Circle the correct choice from the paired terms in parentheses.

1. The period of viability of the sperm is (greater/less) than that for the egg.
2. Sperm mature and acquire the ability to move in the (epididymis/vagina).
3. The fertilized ovum is called a (zygote/morula/blastocyst).
4. (Monozygous/Dizygous) twins are genetically identical.
5. hCG is secreted by (trophoblast/embryoblast) cells.
6. The umbilical (arteries/veins) carry oxygen to the fetus.
7. (hCG/hCS) decreases maternal utilization of glucose to increase the amount available to the fetus.
8. The breasts increase dramatically and begin to secrete colostrum during the (first/second/third) trimester.
9. The fetal heart begins to beat in the (germinal/embryonic/fetal) stage of development.
10. (Prostaglandins/Progesterone) could be used to terminate a pregnancy prematurely.

III. True / False. To the left of the number, write T for true and F for false.

_______ 1. Cervical mucus is most thick and viscous at ovulation.

_______ 2. The acrosomal reaction is not necessary for fertilization to occur.

_______ 3. Monozygous twins result from two sperm fertilizing one ovum.

_______ 4. The ovum completes its second meiotic division just before fertilization.

_______ 5. Fetal hemoglobin has a greater affinity for CO_2 than maternal hemoglobin.

_______ 6. Fetal blood mixes with maternal blood in the placenta.

_______ 7. Gestational diabetes results from an increased sensitivity to insulin during the pregnancy.

_______ 8. The amniotic cavity develops during the fetal stage of development.

_______ 9. Milk is synthesized in the areola of the breast.

_______ 10. Estrogen and progesterone are necessary for milk production and synthesis after parturition.

IV. Sentence Completion. Fill in the blank with the appropriate word.

1. Fertilization occurs in the ____________.
2. The hormone ____________ stimulates the cervix to produce thick, sticky, viscous mucus.
3. The acrosomal reaction occurs when the sperm contacts the ____________ around the ovum.
4. The presence of ____________ is required for the acrosomal reaction and for the electrical block preventing polyspermy.
5. The fetal part of the placenta arises from the ____________ and the maternal part from the ____________.
6. At the end of the first trimester, more progesterone is produced by the ____________ than the ____________.

7. The conversion of cholesterol to androgens occurs in the __________, and the conversion of androgens to estrogen occurs in the __________.

8. __________ and __________ prevent premature expulsion of the fetus.

9. Surfactant would be detected in the __________ month.

10. Cervical dilation stimulates the __________ pituitary to secrete __________.

V. Multiple Choice. Circle the letter of the best response.

1. The sperm's pathway to the egg is:
   a. vagina → cervix → uterus → fallopian tube
   b. uterus → vagina → cervix → fallopian tube
   c. vagina → cervix → fallopian tube → uterus
   d. uterus → cervix → vagina fallopian tube
   e. cervix → vagina → fallopian tube → uterus

2. An ejaculation contains 108 sperm. Only about ________ sperm reach the ovum.
   a. 10^2
   b. 10^3
   c. 10^4
   d. 10^5
   e. 10^6

3. The __________ implants into the uterine endometrium.
   a. zygote
   b. morula
   c. blastocyst
   d. embryoblasts
   e. fertilized ovum

4. The degeneration of the corpus luteum is prevented by:
   a. estrogen
   b. progesterone
   c. hCG
   d. hCS
   e. prolactin

5. A woman has learned that she is pregnant and has to have her ovaries removed. She wants the pregnancy to continue. The pregnancy would be minimally affected if her ovaries were removed:
   a. during the embryonic stage
   b. during the germinal stage
   c. during the first trimester
   d. during the second trimester
   e. her ovaries are necessary throughout the pregnancy

6. hCG has all of the following effects except:
   a. development of maleness in a fetus
   b. maintenance of the corpus luteum
   c. stimulation of uterine contractions
   d. decrease of maternal utilization of glucose
   e. c and d

7. During the pregnancy, all of the hormones are at elevated levels except:
   a. GnRH
   b. estrogen
   c. progesterone
   d. prolactin
   e. hCS

8. At parturition, the hormone present in the lowest quantity is:
   a. prolactin
   b. estrogen
   c. progesterone
   d. oxytocin
   e. prostaglandin

9. Uterine contractions are stimulated by:
   a. progesterone
   b. estrogen
   c. relaxin
   d. oxytocin
   e. c and d

10. Milk let-down is directly triggered by:
    a. prolactin
    b. oxytocin
    c. GnRH
    d. suckling
    e. a baby's crying

VI. Complete the Table.

For each hormone, indicate the source, target tissue, and effect.

Hormone	Source (s)	Target Tissue (s)	Effect
1. hCG			
2. hCS			
3. Estrogen			
4. Progesterone			
5. Oxytocin			
6. Relaxin			
7. Prostaglandin			

VII. Complete the Table.

For each condition, indicate the effect (stimulation, inhibition, none) on the production of the other hormones.

Condition	GnRH	FSH	LH	Testosterone	Estrogen
1. Excess inhibin					
2. Excess testosterone					
3. Decrease in GnRH					
4. Increase in FSH					
5. Increase in LH					
6. Increase in granulosa cells					

Chapter 33

Sexual Physiology

Chapter Learning Objectives

1. Account for sex determination by chromosomes.
2. Trace the early embryonic development of the male and female reproductive systems.
3. Describe abnormal patterns of sexual differentiation.
4. Describe the process of sexual arousal tracing development through the four stages in both sexes.
5. Describe sexual dysfunctions of the male and female.
6. Compare the different methods of contraception as to efficacy, method of operation, and ease of use.

Expanded Chapter Outline

I. Sexual Determination and Differentiation

A. *The chromosomal sex of the individual is established when the sperm and ovum fuse.*

1. Every cell in the body except for the sperm and ovum contains 23 pairs of chromosomes.

2. During meiosis, each sperm and ovum receives one sex chromosome in addition to 22 chromosomes called autosomes.
   a. The sperm receives an X or Y, and each ovum receives an X sex chromosome.
   b. The sex chromosomes are designated **XX** in the female and **XY** in the male.

3. As a result of fertilization, the zygote contains 46 chromosomes and its sex is determined.

B. *The sex chromosomes provide the genetic information that directs differentiation of the gonads into ovaries or testes.*

1. The **X chromosome** carries genes whose pattern of expression depends on the sex of the recipient of these genes.

2. The **Y chromosome** carries genes directing the gonad to differentiate as testes.
   a. The absence of the Y results in ovarian development.
   b. The **H-Y antigen**, produced from the SRY gene, is thought to be responsible for inducing testicular differentiation.

3. During the first six weeks of embryonic development, the gonads of a genetically male or female embryo are anatomically and morphologically identical.

4. Two distinct ductal systems develop in the embryo regardless of genetic/chromosomal sex.
   a. In the female, the **Müllerian duct system** develops into the uterus, fallopian tubes, and vagina, and the Wolffian duct system degenerates.
   b. In the male, the Müllerian duct system regresses, and the **Wolffian duct system** develops into the epididymis, vas deferens, seminal vesicles, prostate, and ejaculatory duct.

C. *Differentiation of the internal and external genitalia is under hormonal control.*

1. Differentiation of the testes begins at 6 to 7 weeks of gestation.
   a. The Sertoli cells secrete **anti-Müllerian hormone (AMH)**, which induces regression of the Müllerian duct system through paracrine action.
   b. The Leydig cells secrete testosterone at 7 to 9 weeks of gestation, which stimulates differentiation of the Wolffian duct through a paracrine action.

2. Differentiation of the ovaries begins about 12 weeks of gestation.
   a. Differentiation of the Mullerian duct does not require the presence of hormones.
   b. The Wolffian duct system regresses.

3. The undifferentiated external genitalia consist of:
   a. An **urogenital slit** surrounded by urethral folds.
   b. **Labioscrotal swellings**.
   c. A **genital tubercle** above the urogenital slit.

4. Differentiation of external genitalia in the female occurs in the absence of hormones.
   a. The urethral folds and labioscrotal swellings become the labia minora and majora, respectively.
   b. The tubercle becomes the clitoris.
   c. The urogenital slit becomes the vaginal opening and urethra.

5. Differentiation of external genitalia in the male occurs in response to dihydrotestosterone (DHT).
   a. External genitalia convert testosterone to DHT by 5α-reductase.
   b. The urethral folds fuse around the urethra to form the shaft of the penis.
   c. The labioscrotal swellings form the scrotum.
   d. The genital tubercle becomes the penis.

D. *Abnormalities of sexual differentiation can occur at several levels.*

1. Normal patterns of sexual differentiation can be altered because of abnormalities at the level of:
   a. Chromosomal sex (XX, XY, genotype).
   b. Gonadal sex (testes, ovaries).
   c. Internal genitalia sex.
   d. External genitalia sex (phenotype).

2. Both X chromosomes in the female are necessary for development of the ovaries, but only one X is needed for differentiation of the Müllerian duct system.
   a. **Turner's syndrome** results when one X is present.
   b. In Turner's syndrome, the gonad develops into a vestigial gonadal streak, but the fallopian tubes, uterus, and external genitalia develop normally.

3. The presence of an extra X in the male results in **Klinefelter syndrome**, which is characterized by testicular dysfunction, impaired development of the seminiferous tubules, and a deficiency of spermatogenesis.

4. Exposure of a genetically female embryo to androgens can result in masculinization, particularly of external genitalia.
   a. These females lack 11- or 21-hydroxylase to synthesize glucose and mineralocorticoids.
   b. Cholesterol is converted to androgens whose hypersecretion leads to development of **adrenogenital syndrome**.

5. Genetic males who lack androgen receptors cannot respond to testosterone, which results in a condition known as **testicular feminization syndrome**.
   a. The testes develop, but the Wolffian duct system does not.
   b. Deficiency of enzymes necessary for testosterone biosynthesis results in reduced masculinization of external genitalia.

II. Human Sexual Response

A. *The physiology of the sexual response in both males and females can be divided into four stages: excitement, plateau, orgasm, and resolution.*

1. Sexual arousal can result from physical or mental stimulation, but the physiological responses to it are the same.

2. Physiological responses occurring in the **excitement** phase in the **male** include:
   a. Erection of the penis from vasocongestion.
   b. Increased blood flow to the testes and subsequent increase in testicular size.
   c. An increase in general muscle tone or myotonia.

3. Physiological responses occurring in the excitement phase of females include:
   a. Increased blood flow to the vagina and external genitalia, resulting in increased size of the clitoris and both labia.
   b. Vaginal lubrication from vasocongestion.
   c. Expansion of the vagina.
   d. Lifting of the uterus and cervix.
   e. Increased myotonia.

4. Physiological responses in the **plateau phase** of **males** include:
   a. Secretion (of pre-ejaculatory fluid) from bulbourethral glands into the urethra.
   b. Increased heart rate, blood pressure, respiration rate, and myotonia.

5. **Plateau phase** physiological responses in the **female** include:
   a. Tenting of the uterus.
   b. Retraction of the clitoris.
   c. Increased heart and respiration rates, increased blood pressure, and myotonia.

6. **Ejaculation** during the **orgasmic phase** in the **male** involves two processes:
   a. **Emission**, in which ejaculatory fluid is forced into the urethra by rhythmic muscular contractions.
   b. **Expulsion**, in which rhythmic muscular contractions of the urethra force semen out the penile tip.

7. **Orgasmic phase** in the **female** is characterized by rhythmic muscular contractions of the vagina and uterus and other muscles. Orgasm is not essential for reproduction.

8. During **resolution**, the physiological responses return to pre-excitement levels in the **male**.
   a. The phase is characterized by loss of penile erection, decrease in testicular size, muscle relaxation, and return to pre-excitement physiological levels.
   b. The length of the **refractory period** is highly variable and is affected by age, fatigue, and time since the last orgasm.

9. During **resolution**, the physiological responses return to pre-excitement levels in the **female**.
   a. Resolution is characterized by muscle relaxation and decrease in blood flow to breasts, clitoris, labia, and vagina.
   b. Females are multiorgasmic, unlike males.

10. Sexual arousal in response to physical stimulation occurs even before birth.

B. *Changes occur in the male and female sexual response during aging.*

1. In the female, the breasts, clitoris, and vagina remain sensitive to stimulation in old age.

2. Physiological changes in the female during aging include:
   a. Decreased blood flow to the vagina resulting in decreased vaginal lubrication.
   b. Decrease in amount of myotonia previously occurring during excitement.
   c. Decrease in vaginal elasticity.

3. In the male, sexual response physiological changes during aging include:
   a. Decreased sperm production.
   b. Increase in time spent in refractory period.
   c. Increase in amount of time and stimulation required for erection.
   d. Decrease in volume of ejaculatory fluid.

III. Sexual Dysfunction

A. *Sexual dysfunction refers to any condition in which normal physical responses to sexual stimulation are impaired.*

1. An estrogen deficiency in the female may result in **atrophic vaginitis**, which is characterized by dryness and thinning of vaginal mucosa, decreased elasticity, and decreased muscle tone of the vagina.

2. Testosterone deficiency in the male negatively affects sexual desire and ejaculation.

3. Mechanical and disease-related injuries to the spinal cord or brain that interrupt afferent and efferent pathways can result in impairment of sexual arousal and orgasm in both males and females.
   a. Multiple sclerosis may be accompanied by a reduction or elimination of sexual arousal and orgasmic capacity.
   b. Neural dysfunction may result from diabetes.

4. Large amounts of alcohol, sedatives, narcotics, and tranquilizers can retard or inhibit sexual arousal and orgasm in both males and females.
   a. Other drugs (antihypertensives) can alter neural control of blood flow to genitalia.
   b. Antihistamines containing ephedrine decrease blood flow to the vagina and vaginal lubrication.

5. Other factors may result in sexual dysfunction in females.
   a. **Anorgasmia**, difficulty in reaching an orgasm, is due primarily to psychological factors.
   b. **Nonatrophic vaginitis** can be caused by infection, allergic reactions, or chemical irritation.

c. **Endometriosis** and **pelvic inflammatory disease** can cause pain during arousal and intercourse.
d. **Vaginismus**, or involuntary contractions and spasms of the vaginal muscles, makes penetration difficult.

6. Other factors may result in sexual dysfunction in the male.
   a. **Erectile dysfunction** or **impotence**, defined as the inability to achieve penile erection sufficient for vaginal penetration, may be classified as primary or secondary.
   b. Secondary dysfunction can occur in response to anxiety, illness, fatigue, lack of privacy, or alcohol consumption.
   c. Premature ejaculation occurs more often in younger males and after prolonged stimulation before vaginal penetration.
   d. **Dyspareunia**, or painful erection, intromission, or ejaculation, can result from **balanitis** or **urethritis**, penile scarring, or reduced elasticity of the skin covering the penis.

B. *Infertility, or the inability to reproduce, can be due to anatomical, hormonal, chromosomal, immunological, or psychological factors.*

1. Hormonal factors include:
   a. Inadequate secretion of gonadotropins, androgens, or female sex hormones.
   b. Lack of response or receptors in target tissues.

2. **Infertility** may result from low sperm count, abnormal sperm shape or structure, and decreased motility.

3. Immunological factors include the production of antisperm antibodies in both males and females.

IV. Contraceptive Methods

A. *Barrier methods of contraception prevent sperm transport.*

1. Barrier methods include:
   a. Condoms.
   b. Diaphragms and cervical caps.
   c. Contraceptive sponges.

2. They vary in effectiveness and have minimal side effects.

B. *Chemical methods of contraception include birth control pills, which contain synthetic variations of gonadal hormones.*

1. Chemical methods include:
   a. Spermicides.
   b. Birth control pill, which blocks ovulation by interfering with the release of FSH and LH from the pituitary.
   c. Morning-after pill, which contains a high level of estrogen and prevents implantation.

2. Birth control pills are effective but may produce a variety of side effects.

3. The intrauterine device is believed to interfere with implantation.

C. *Surgical methods of contraception include vasectomy in the male and tubal ligation in the female.*

1. These methods are relatively safe, effective, and permanent.

2. Postsurgical complications may include bleeding and infection.

D. *Abstinence is the only completely effective form of birth control.*

1. Periodic abstinence during the time of ovulation is termed the rhythm method.

2. The efficacy of this method depends on the accuracy of predicting ovulation.

3. Three methods to predict ovulation involve monitoring the:
   a. Ovarian cycle and length of the menstrual cycle.
   b. Daily basal body temperature.
   c. Consistency of cervical mucus.

E. *Several new approaches to contraception are currently being developed and tested.*

1. Important variables to consider in the development of new contraceptives are efficacy, duration of action, reversibility, ease of use, and incidence of short-term and long-term side effects.

2. Research efforts for the female contraceptive include:

a. Development of long-acting forms of estrogen and progesterone, which can be administered by subcutaneous implantation or injection.
b. Antagonists to gonadotropins.
c. Synthetic analogs of prostaglandins.

3. In the male, research is focused in blocking production, maturation, motility, or transport of sperm.

Key Terms

abstinence
adrenogenital syndrome
amenorrhea
anorgasmia
arousal
atrophic vaginitis
autosomes
birth control pill
contraception
dihydrotestosterone
dyspareunia
emission
endometriosis
erectile dysfunction
erection
excitement
expulsion
genetic sex (XX, XY)
genital tubercle
gonadal sex
H-Y antigen
infertility
intrauterine device
Klinefelter syndrome
labioscrotal swelling
masculinization
meiosis
morning after pill
Müllerian duct system
Müllerian inhibiting substance
myotonia
nonatrophic vaginitis
orgasm
pelvic inflammatory disease
phenotypic sex
plateau
premature ejaculation
primitive gonads
resolution rhythm method
sex chromosomes
sexual determination
sexual differentiation
sexual dysfunction
testicular feminization
testosterone
tubal ligation
Turner syndrome
urogenital folds, slit
vaginal lubrication
vaginismus
vasectomy
vasocongestion
Wolffian duct systems

Testing Your Knowledge

I. Matching. Select the correct term from the Key Terms.

_________ 1. sex chromosomes in female

_________ 2. develops into clitoris or penis

_________ 3. responsible for inducing testicular differentiation

_________ 4. develops into uterus, vagina, cervix

_________ 5. develops into vas deferens, epididymis

_________ 6. becomes scrotum and labia majora

_________ 7. stimulates differentiation of external genitalia

_________ 8. abnormality in male with extra X chromosome

_________ 9. syndrome with masculinization of external genitalia

_________ 10. syndrome with absence of masculinization in external genitalia

_________ 11. sexual response stage characterized by erection and vaginal lubrication

_________ 12. sexual response stage characterized by ejaculation and decrease in myotonia

_________ 13. dysfunction resulting from inadequate estrogen stimulation

_________ 14. involuntary contractions and spasms of pubococcygeal muscles

_________ 15. surgical procedure in which male transport duct is severed

II. Dichotomous. Circle the correct choice from the paired terms in parentheses.

1. The genetic sex is determined by the (sperm/ovum).

2. In the female the (Mullerian/Wolffian) duct system differentiates into fallopian tubes, uterus, cervix, and vagina.

3. Differentiation of the testes occurs (earlier/later) than that of the ovaries.

4. Müllerian duct development (requires/does not require) the presence of estrogen.

5. A genetic male with a deficiency of androgen receptors would have (male/female/ambiguous) external genitalia.

6. A genotype of XXY is associated with (Turner syndrome/Klinefelter syndrome/testicular feminization).

7. Vaginal lubrication and penile erection occur in response to (vasodilation/vasocongestion/vasoconstriction).

8. Emission and expulsion occur during the (plateau/orgasm/resolution) phase of sexual arousal.

9. Orgasm involves increased activity of (efferent/afferent) neurons.

10. A low sperm count may result in (infertility/impotence).

III. True/False. In the space to the left of the number write T for true or F for false.

_______ 1. In the female, just one X chromosome is necessary for the development of the ovaries.

_______ 2. Removal of the right testes in an eight-week embryo would have no effect on the formation of the vas deferens on the right side.

_______ 3. Testosterone stimulates sexual development of the penis and scrotum.

_______ 4. The scrotum and labia majora have a common embryological origin.

_______ 5. Males are multiorgasmic.

_______ 6. Rhythmic muscular contractions occur during the orgasmic phase.

_______ 7. A woman whose ovaries have been removed is incapable of having a sexual response.

_______ 8. Testosterone is necessary for male sexual response.

_______ 9. The sexual responses of a 70-year-old and a 17-year-old woman are the same.

_______ 10. Impotence and infertility are synonymous terms.

IV. Sentence Completion. Write your response in the space provided.

1. The __________ carries genetic loci, which I are required for male development.

2. The __________ cells of the testes secrete AMH and the __________ cells secrete testosterone.

3. __________ syndrome is characterized by streak gonads and normal external and internal genitalia.

4. The penis and __________ have a common embryological origin.

5. Significant increases in heart rates, blood pressure, and respiration rate occur during the __________ phase.

6. Many changes in sexual responses that occur in an aging female result from a decrease in __________.

7. The __________ stage is most similar in a 60-year-old male and a 16-year-old male.

8. Anorgasmia is due primarily to __________ factors.

9. Continuous presence of estrogen and progesterone inhibit the release of __________ and __________ from the __________.

10. The morning-after pill contains a high level of __________ and acts primarily by preventing __________.

V. Multiple Choice. Circle the letter of the best response.

1. All of the following pairs are correctly matched except:
   a. Wolffian duct → epididymis
   b. Müllerian duct → uterus
   c. genital tubercle → clitoris
   d. urethral fold → labia minora
   e. urogenital sinus → ejaculatory duct

2. In the development of the male, the first event to occur is:
   a. differentiation of the testes
   b. regression of the Müllerian duct
   c. differentiation of Wolffian duct
   d. fusion of urethral folds to form penile shaft
   e. fusion of labioscrotal swellings to form scrotum

3. In the development of the female, the first event to occur is:
   a. differentiation of the Müllerian duct

b. degeneration of the Wolffian duct
c. differentiation of the ovary
d. differentiation of the labioscrotal swellings into labia majora
e. differentiation of the genital tubercle into the clitoris

4. A person with testes, no vas deferens or epididymis, and a clitoris and labia would have:
   a. Turner syndrome
   b. Klinefelter syndrome
   c. adrenogenital syndrome
   d. testicular feminization
   e. 5 α-reductase deficiency

5. A female with adrenal 11- or 21-hydroxylase deficiency would have:
   a. ovaries, female internal and external genitalia
   b. streak gonads, female internal and external genitalia
   c. testes, no internal genitalia, female external genitalia
   d. testes, male internal genitalia, ambiguous external genitalia
   e. ovaries, female internal genitalia, ambiguous external genitalia.

6. Vasocongestion and myotonia first occur during the _______ phase.
   a. excitement
   b. plateau
   c. orgasm
   d. resolution

7. The two stages of the human sexual response that are most different between males and females in terms of physical and physiological changes are:
   a. orgasm and resolution
   b. excitement and plateau
   c. excitement and orgasm
   d. plateau and resolution
   e. excitement and resolution

8. Which of the following events would not occur in a 20-year-old man whose efferent neurons to blood vessels and muscles of internal and external genitalia were severed?
   a. penile erection
   b. emission
   c. expulsion
   d. movement of the testes closer to the body
   e. a-d would not occur

9. The most common type of sexual dysfunction in a female is:
   a. anorgasmia
   b. vaginismus
   c. atrophic vaginitis
   d. dyspareunia
   e. nonatrophic vaginitis

10. The most effective method of contraception is (the):
    a. condom
    b. birth control pill
    c. tubal ligation vasectomy
    d. periodic abstinence
    e. diaphragm

VI. Complete the Table.

Embryonic Structure	Adult Structure Male	Adult Structure Female
1. Genital tubercle		
2. Labioscrotal swellings		
3. Urogenital folds		
4. Urogenital slit		
5. Wolffian duct system		
6. Gonad		

VII. Complete the Table.

For each syndrome, indicate the cause and appearance of structure (if known).

Syndrome	Cause	External Genitalia	Gonad	Internal genitalia
Turner				
Klinefelter				
Adrenogenital				
Testicular feminization				

VIII. Complete the Table.

For each condition describe the effect on sexual differentiation for the stated sex.

Embryonic Structure	Adult Structure Female
1. Exposure of male embryo to testosterone	
2. Exposure of male embryo to estrogen and progesterone	
3. Deficiency of 5 α-reductase in a male embryo	
4. Deficiency of androgen receptors in a male embryo	
5. Embryo with XX	
6. Deficiency of 21-hydroxylase in a female embryo	

Answers

CHAPTER 1

I. Matching

1. deductive reasoning
2. inductive reasoning
3. Harvey
4. Bernard
5. ventral
6. hypothesis
7. sagittal
8. ratio
9. inverse
10. proximal
11. tissue
12. connective
13. interstitial fluid
14. homeostasis
15. sensor or effector

II. Calculations

1. a. 10^{-3} gm
   b. 10^{-6} gm
   c. 10^{-9} gm
   d. 10^{-12} gm

III. Fill in the Table

Height	Weight
1.63×10^{-3} km	5.45×10^{1} kg
$1.63 \times$ m	5.45×10^{4} gm
1.63×10^{2} cm	5.45×10^{7} mg
1.63×10^{3} mm	5.45×10^{10} µg
1.63×10^{6} µm	5.45×10^{13} ng
1.63×10^{9} nm	5.45×10^{16} pg

IV. Interpretation of Graphs

a. inverse
b. 25 bubbles
c. distance
d. number of bubbles
e. set-up in which the light is stationary

CHAPTER 2

I. Matching

1. buffer
2. exergonic
3. cation
4. steroid
5. unsaturated
6. energy of activation
7. atomic number
8. alpha helix, beta sheet
9. glycoprotein
10. hydrogen
11. acid
12. peptide

II. Dichotomous

1. mass
2. anions
3. positively
4. loses
5. Hydrogen
6. hydrophobic
7. low
8. deoxyribose
9. cofactor
10. Exergonic

III. True / False

True: 1, 2, 4
False: 3, 5, 6, 7, 8, 9, and 10

IV. Multiple Choice

1.c 2.d 3.d 4.b 5.b 6.c 7.a 8.b 9.c 10.b

V. Visual # 1

a. 4
c. 6
b. 6
d. 6
e. covalent

CHAPTER 3

I. Matching

1. chromatin
2. nuclear envelope
3. mitochondria
4. interphase
5. ribosome
6. rough ER
7. nucleolus
8. prokaryote
9. lysosome
10. exocytosis
11. endocytosis
12. smooth ER
13. Golgi Complex
14. neoplasm
15. cytoskeleton

II. Dichotomous

1. interphase
2. peroxisomes
3. exocytosis
4. rough ER
5. Many
6. cytoplasm
7. identical
8. antigen
9. mitochondria
10. one

III. True/False

True: 3, 6, 9, and 10
False: 1, 2, 4, 5, 7, and 8

IV. Sentence Completion

1. eukaryote
2. gene
3. Rough ER and Golgi
4. Golgi
5. Endocytosis
6. Exocytosis
7. genome
8. lipids, steroids
9. microfilaments
10. Radioimmunoassay,

V. Multiple Choice

1.b 2.b 3.a 4.c 5.c 6.a 7.c 8.c 9.b 10.b

VI. Problem

a. cytochemistry, light, and electron microscopy
b. #2
c. autoradiography
d. in all pellets
e. #1
f. #3

CHAPTER 4

I. Matching

1. diffusion
2. active transport
3. amphipathic
4. facilitated diffusion
5. asymmetry
6. symport
7. extrinsic
8. glycoprotein
9. osmosis
10. intrinsic
11. Na^+-K^+ ATPase pump
12. lipid bilayer
13. cholesterol
14. glycolipid
15. Van't Hoff law

II. Dichotomous

1. Intrinsic
2. Proteins
3. greater
4. decreases
5. Active
6. increases
7. lower
8. increases
9. basolateral
10. out of

III. True/False

True: 1, 5, 8, and 9
False: 2, 3, 4, 6, 7, and 10

IV. Sentence Completion

1. intercellular or between cell
2. microvilli
3. lateral, flip-flop
4. olive oil
5. mitochondria
6. Carbohydrates
7. Fick's law
8. passive
9. osmolarity
10. active

V. Multiple Choice

1.e 2.e 3.d 4.e 5.b 6.a 7.b 8.e 9.c 10.e

VI. Problem Solving

1a. from the bag to the beaker
1b. hyperosmotic
1c. hypo-osmotic
2a. from the bag to the beaker
2b. hyperosmotic
2c. hypo-osmotic

CHAPTER 5

I. Matching

1. triplet
2. anticodon
3. translation
4. agonist
5. RNA polymerase
6. DNA polymerase
7. DNA ligase
8. transcription
9. tRNA
10. mRNA
11. cyclic AMP
12. coupling
13. receptor
14. codon
15. lagging strand

II. Dichotomous

1. codon
2. transcription
3. thymine
4. introns
5. A site
6. plasma membrane
7. cytoplasmic
8. agonist
9. second
10. calcium

III. True/False

True: 2, 3, 7, and 8
False: 1, 4, 5, 6, 9, and 10

IV. Sentence Completion

1. methionine
2. aminoacyl synthetase
3. gap junctions
4. water
5. 20, 64
6. tRNA, mRNA
7. receptors
8. polysomes
9. tRNA
10. gap junction

V. Multiple Choice

1.a 2.a 3.c 4.d 5.e 6.a 7.d 8.d 9.a 10.e

VI. Complete the Table

DNA	TAC	AGC	ACA	CAT	ATC
mRNA	AUG	UCG	UGU	GUA	UAG
tRNA	UAC	AGC	ACA	CAU	—
amino acid	met	ser	cys	val	—

CHAPTER 6

I. Matching

1. allosteric inhibition
2. glycolysis
3. citric acid cycle
4. reduction
5. oxidation
6. substrate level phosphorylation
7. redox potential
8. mitochondrial matrix
9. cytochrome
10. oxaloacetate
11. deamination
12. acetyl group
13. NADPH
14. ATP synthetase
15. cellular respiration

II. Dichotomous

1. catabolism
2. positive
3. NADH
4. low
5. mitochondria
6. accelerated
7. reduced
8. $NADP^+$
9. less
10. can

III. True / False

True: 2, 3, 5, 6, 7, 9 and 10
False: 1, 4, and 8

IV. Sentence Completion

1. Photosynthesis, cellular respiration
2. oxidation, reduction
3. low, high
4. negative
5. acetyl
6. mitochondrial matrix
7. ten, two
8. three
9. mitochondrial matrix, intermembrane space
10. mitochondrial matrix, cytosol

V. Multiple Choice

1.b 2.e 3.e 4.e 5.d 6.a 7.a 8.e 9.b 10.d

CHAPTER 7

I. Matching

1. activating state
2. absolute refractory period
3. saltatory conduction
4. electrical synapse potential
5. excitatory post-synaptic
6. glia
7. IPSP
8. soma
9. sensory neuron
10. acetylcholine
11. synaptic vesicle
12. discrete synapse
13. dopamine
14. neuropeptide
15. spatial summation

II. Dichotomous

1. Motor
2. Neurons
3. dendrites
4. smooth
5. inversely
6. K^+
7. equals
8. absolute
9. electrical
10. more

III. True / False

True: 5, 9, 10
False: 1, 2, 3, 4, 6, 7, 8

IV. Sentence Completion

1. peripheral
2. bipola
3. axon terminals
4. Fast
5. Chemically activated
6. depolarization
7. Na^+, K^+
8. greater
9. Ca^{2+}
10. muscarinic

V. Multiple Choice

1.e 2.c 3.a 4.b 5.c 6.d 7.a 8.e 9.c 10.d

VI. Label the Diagram

1. resting stage
2. threshold, depolarization
3. rising phase, depolarization
4. repolarizing phase, hyperpolarization
5. absolute refractory period, hyperpolarization

CHAPTER 8

I. Matching

1. dorsal root ganglion
2. dorsal horn
3. generator potential
4. thalamus
5. frequency code of stimulus intensity
6. sensory adaptation
7. proprioceptors
8. nocioceptors
9. phasic discharge
10. fovea
11. cones
12. ganglion cell
13. lateral geniculate nucleus
14. hair cell
15. gustatory or olfactory

II. Dichotomous

1. mechano-
2. phasic
3. greatest
4. anterolateral
5. horizontal
6. lateral geniculate
7. independently
8. tectorial
9. amplitude
10. ossicles

III. True / False

True: 1, 6, and 10
False: 2, 3, 4, 5, 7, 8, and 9

IV. Sentence Completion

1. rostrocaudal
2. dorsal
3. neocortex
4. ciliary muscles
5. three
6. basilar
7. high
8. interaural time difference
9. inferior colliculus
10. olfactory

V. Multiple Choice

1.d 2.b 3.c 4.c 5.c 6.a 7.a 8.a 9.a 10.a.

CHAPTER 9

I. Matching

1. motor neurons
2. extrafusal fiber
3. fast skeletal muscle
4. myoneuronal junction
5. alpha motor neuron
6. motor unit
7. spindle fiber
8. Golgi tendon organ
9. stretch reflex
10. inverse myotatic reflex
11. locomotor generator
12. ventral horn
13. Purkinje
14. semicircular canals
15. saccule, utricle

II. Dichotomous

1. Fast
2. one motor neuron and several muscle fibers
3. greater
4. Excitatory
5. muscle spindles
6. Alpha
7. greater
8. more
9. neocortex
10. Angular

III. True / False

True: 5 and 7
False: 2, 3, 4, 6, 8, 9, and 10

IV. Sentence Completion

1. extensor
2. ventral
3. interneurons
4. stance
5. increases
6. pyramidal
7. somatic sensory
8. posterior parietal
9. Purkinje
10. basal ganglia, substantia nigra

V. Multiple Choice

1.e 2.e 3.e 4.a 5.b 6.e 7.c 8.c 9.d 10.e

CHAPTER 10

I. Matching

1. acetylcholine
2. interomedial lateral nuclei
3. muscarinic, nicotinic
4. alpha, beta
5. continence
6. osmoreceptors
7. ADH, vasopressin
8. carotid, aortic bodies
9. medulla
10. baroreceptors
11. depressor center
12. pressor center
13. adrenal medulla
14. CSF
15. norepinephrine

II. Dichotomous

1. short
2. ventral
3. cholinergic
4. hyperpolarization
5. inhibits
6. Antidiuretic hormone
7. Aortic bodies
8. more
9. depressor
10. cortex

III. True / False

True: 4, 5, 6, 7, 8, and 10
False: 1, 2, 3, and 9

IV. Sentence Completion

1. thoracic, lumbar
2. Acetylcholine
3. increases
4. micturition
5. Hyperalgesic zones
6. supraoptic, paravent-icular, hypothalamus
7. kidney, decrease
8. decreases, decreases
9. increase
10. corticotropin releasing hormone (CRH)

V. Multiple Choice

1.a 2.b 3.e 4.e 5.a 6.d 7.d 8.a 9.a 10.b

CHAPTER 11

I. Matching

1. superchiasmatic nucleus
2. electroencephalo-gram
3. REM
4. preoptic area
5. raphe nucleus
6. anterior hypothal-amus
7. supraoptic nucleus
8. androgen
9. substantia nigra hypothalamus
10. insomnia
11. habituation
12. classical conditioning
13. Wernicke's area
14. Broca's area
15. aphasia

II. Dichotomous

1. decreases, increases
2. parasympathetic
3. glucose
4. supraoptic
5. anterior
6. androgens
7. learning
8. long-term
9. hippocampus
10. left

III. True / False

True: 2, 4, and 8
False: 1, 3, 5, 6, 7, 9, and 10

IV. Sentence Completion

1. diurnal or circadian
2. suprachiasmatic nucleus
3. hypothalamus
4. supraoptic nucleus, paraventricular nucleusADH or vasopressin
5. Pheromones
6. Sensitization
7. consolidation
8. classical
9. operant
10. Wernicke's area, Broca's area

V. Multiple Choice

1.c 2.a 3.e 4.e 5.c 6.a 7.a 8.a 9.c 10.a

CHAPTER 12

I. Matching

1. steroid
2. albumin, prealbumin
3. amine
4. thyroxine
5. peptide hormones
6. dynamic equilibrium
7. receptor- mediated endocytosis
8. cAMP
9. IP_3, DG
10. IP_3
11. steroid
12. mineralocorticoid
13. insulin, glucagon
14. cholesterol
15. signal transducer

II. Dichotomous

1. Paracrine
2. thyroxine
3. cortex, medulla
4. free
5. low
6. intracellular
7. reversible
8. Rs
9. raised
10. less

III. True / False

True: 4, 8, and 10
False: 1, 2, 3, 5, 6, 7, and 9

IV. Sentence Completion

1. amines
2. thyroglobulin
3. cholesterol
4. pregnenolone
5. liver
6. liver, kidney
7. competitive binding
8. cell surface
9. signal transducers
10. amplification

V. Multiple Choice

1.d 2.e 3.b 4.a 5.e 6.e 7.e 8.d 9.c 10.c

CHAPTER 13

I. Matching

1. anterior pituitary
2. posterior pituitary
3. supraoptic paraventricular nuclei system
4. hypothalamus-pituitary portal
5. somatostatin
6. ADH
7. oxytocin
8. FSH, LH
9. growth hormone
10. somatomedins, IGF-I, -II
11. IGF-I
12. dwarfism
13. thyroglobulin
14. TSH, TRH
15. myxedema

II. Dichotomous

1. anterior
2. posterior, axons
3. increase
4. increases
5. PRF
6. anabolic
7. Somatostatin
8. decreased
9. basal
10. TSH

III. True / False

True: 2, 4, 7, and 8
False: 1, 3, 5, 6, 9, and 10

IV. Sentence Completion

1. anterior, hypothalamic- pituitary
2. ADH
3. breast, uterus
4. oxytocin, prolactin
5. somatotrophs
6. CRH
7. chondrocytes, epiphyseal plates
8. tyrosine
9. negative
10. four, three

V. Multiple Choice

1.a 2.b 3.b 4.b 5.e 6.d 7.e 8.d 9.c 10.c

VI. Fill in the Blank

1. decrease
2. increase
3. decrease
4. no effect
5. decrease
6. decrease

VII. Fill in the Blank

1. decrease
2. decrease
3. no effect
4. no effect
5. decrease
6. increase
7. decrease
8. decrease
9. increase

CHAPTER 14

I. Matching

1. zona fasciculata
2. zona glomerulosa
3. zona reticularis
4. chromaffin cell, adrenal medulla
5. Cushing's syndrome
6. Addison's disease
7. ACTH, cortisol
8. anti-inflammatory
9. cortisol, glucocorticoid
10. renin
11. angiotensin II
12. aldosterone
13. epinephrine, norepinephrine
14. epinephrine
15. corticosteroid-binding, globulin, immunosuppressive

II. Dichotomous

1. fasciculata
2. cortex
3. increase
4. decreases
5. inside
6. catabolic
7. increase
8. opposite to
9. increase, lowers
10. increased

III. True / False

True: 3, 6, and 10
False: 1, 2, 4, 5, 7, 8, and 9

IV. Sentence Completion

1. CRH, ACTH, cortex
2. cholesterol, pregnenolone
3. 24
4. gluconeogenesis
5. decreased
6. vascular reactivity
7. lymphocytes, eosinophils, basophils; platelets, RBCs, neutrophils
8. potassium ions, angiotensin II
9. excess, deficiency
10. epinephrine

V. Multiple Choice

1.e 2.a 3.b 4.e 5.a 6.e 7.c 8.d 9.a 10.b

VI. Fill in the Blank

1. decrease
2. no effect
3. no effect
4. increase
5. decrease
6. decrease
7. increase

VII. Fill in the Blank

1. decrease
2. no effect
3. no effect
4. decrease
5. decrease
6. decrease
7. decrease

VIII. Fill in the Blank

1. decrease
2. increase
3. increase
4. no effect
5. no effect

CHAPTER 15

I. Matching

1. alpha
2. beta
3. islets of Langerhans
4. proinsulin
5. glucagon
6. insulin, somatostatin
7. anabolic
8. glycogenolysis
9. gluconeogenesis
10. Type I IDDM
11. Type II NIDDM
12. hyperglycemia
13. electrolyte imbalance
14. diabetic neuropathy
15. delta

II. Dichotomous

1. endocrine
2. increases
3. Epinephrine
4. alpha
5. glucagon
6. anabolic
7. increases
8. glycogenolysis
9. insulin
10. increase

III. True / False

True: 1, 3, 6, 7, and 8
False: 2, 4, 5, 9, and 10

IV. Sentence Completion

1. Insulin
2. C-peptide
3. CNS
4. 75-115 mg
5. lipoprotein lipase
6. increases, decreases
7. liver, catabolic
8. ketones
9. high
10. triglycerides

V. Multiple Choice

1.d 2.d 3.a 4.c 5.d 6.a 7.c 8.d 9.a 10.a 11.d

VI. Fill in the Blank

1. decrease
2. increase
3. decrease
4. no direct effect
5. decrease
6. increase

VII. Fill in the Blank

1. decrease
2. increase
3. increase
4. increase
5. increase
6. no direct effect

CHAPTER 16

I. Matching

1. tendon
2. antagonistic pair
3. sarcomere
4. actin
5. myosin
6. sarcolemma
7. sarcoplasmic reticulum
8. intercalated disks, gap junctions
9. isometric
10. isotonic
11. power output
12. t-system
13. tetanus
14. lactic acid
15. white fiber

II. Dichotomous

1. ANS, CNS
2. Unitary
3. Myosin, A band
4. I band, A band
5. isometric
6. 10-6 mM, 10-2 mM
7. Tropomyosin
8. Actin-linked, myosin-linked
9. Phase
10. Atrophy

III. True / False

True: 1 and 4
False: 2, 3, 5, 6, 7, 8, 9, and 10

IV. Sentence Completion

1. skeletal
2. skeletal, bones
3. sarcomere, skeletal
4. six
5. t-tubule system, sarcoplasmic reticulum
6. two
7. calmodulin, troponin,
8. tetanus, summation
9. cardiac, smooth (unitary), skeletal
10. anaerobic (glycolytic)

V. Multiple Choice

1.d 2.b 3.e 4.c 5.c 6.b 7.b 8.e 9.d 10.a

VI. Label the Diagram

1. c, thick filament
2. b, thin filament
3. b, thin filament
4. e, I band
5. a, e, H zone, I band
6. d, A band

VII. Fill in the Blank

1. contract
2. not contract - fatigue
3. continual contraction, cannot relax
4. keeps contracting
5. no contraction

CHAPTER 17

I. Matching

1. erythrocyte
2. erythropoiesis
3. viscosity
4. hemoglobin
5. hemolysis
6. anemia
7. neutrophil
8. eosinophil
9. cell-mediated
10. B-lymphocyte, plasma cell
11. leukemia
12. platelet
13. thrombin
14. complement
15. cytotoxic T

II. Dichotomous

1. Hypervolemia
2. Leukocytes
3. Serum
4. hemolyze
5. beta
6. aplastic
7. low
8. neutrophils
9. leukocytosis
10. Platelets
11. Hemophilia

III. True / False

True: 3, 4, 9, and 10
False: 1, 2, 5, 6, 7, and 8

IV. Sentence Completion

1. increased
2. 93, 7
3. 20 mL
4. vitamin B12, intrinsic factor
5. B or O
6. Basophils
7. T-lymphocytes, lymphokines
8. functions
9. K, calcium
10. prothrombin, thrombin, fibrinogen, fibrin
11. TPA, plasminogen, plasmin
12. Thrombocytopenia

V. Multiple Choice

1.a 2.c 3.b 4.b 5.b 6.c 7.e 8.c 9.a 10.c 11.d and e 12.c 13.d 14.c

CHAPTER 18

I. Matching

1. atrium
2. ventricle
3. atrioventricular valve
4. pulmonary valve
5. myocardium
6. coronary vessels
7. sinoatrial node
8. Purkinje fibers
9. ECG
10. diastole
11. systole
12. diastole
13. stroke volume
14. tachycardia
15. cardiac output

II. Dichotomous

1. unoxygenated
2. inversely
3. tricuspid and bicuspid
4. absolute
5. speeds up, increases
6. greater
7. systole, diastole
8. increases, increases
9. direct
10. aerobic

III. True / False

True: 4, 8, and 10
False: 1, 2, 3, 5, 6, 7, and 9

IV. Sentence Completion

1. right atrium, left atrium
2. diastolic potential
3. sinoatrial node, atrioventricular node
4. reverse, aortic and pulmonary artery, left and right ventricular
5. QSR complex
6. stroke volume, end systolic volume
7. heart rate, stroke volume
8. hyperpolarizes, decreases
9. calcium
10. increase

V. Multiple Choice

1.c 2.c 3.c 4.d 5.b 6.a 7.e 8.c 9.e 10.c 11.e

CHAPTER 19

CHAPTER 19

I. Matching

1. pulse pressure
2. systolic pressure
3. myogenic activity
4. autoregulation
5. vasodilation
6. vasoconstriction
7. vasodilation
8. oncotic pressure
9. Starling's hypothesis
10. filtration
11. absorption
12. veins
13. baroreceptors
14. lymphatic system
15. hypertension

II. Dichotomous

1. arterioles
2. increases
3. decrease
4. alpha, both
5. inversely
6. greater, filtration
7. increase
8. increase, increased
9. increased, decreased
10. vasodilation

III. True / False

True: 2, 4, 5, 8, and 10
False: 1, 3, 6, 7, and 9

IV. Sentence Completion

1. diastole
2. 30 mm Hg
3. decreases
4. pressure, concentration
5. +10 mm Hg, −10 mm Hg
6. medulla
7. total blood volume, redistributing
8. right
9. antidiuretics

V. Multiple Choice

1.a 2.b 3.d 4.a 5.c 6.c 7.e 8.a 9.e 10.e

VI. Complete the Table

Vasodilation: 1, 2, 3, 4, 5, 6, and 8
Vasoconstriction: 5 and 7

VII. Complete the Table

Arterial pressure increases: 1, 3, 4, and 6
Arterial pressure decreases: 2, 5, 7, and 8
Cardiac output increases: 1, 3, 4, 6, and 8
Cardiac output decreases: 2, 5, and 7

CHAPTER 20

I. Matching

1. alveolus
2. ventilation
3. turbulent air flow
4. tidal volume
5. forced vital capacity
6. residual volume
7. functional residual capacity
8. dead space volume
9. diaphragm
10. transpulmonic
11. pneumothorax
12. alveolar
13. thoracic cavity
14. compliance
15. pulmonary surfactant

II. Dichotomous

1. inverse
2. more, less
3. pleural
4. forced
5. directly
6. greater
7. residual
8. Emphysema
9. emphysema
10. smaller

III. True / False

True: 2, 4, 5, and 10
False: 1, 3, 6, 7, 8, and 9

IV. Sentence Completion

1. less
2. −7 cm H_2O
3. 7 cm H_2O
4. restrictive
5. obstructive
6. 16
7. decreases
8. 9000 mL/minute
9. 4.9 L
10. 6000 mL or 6 L

V. Multiple Choice

1.e 2.c 3.d 4.d 5.c 6.d, 7.d, 8.b, 9.b, 10.c

VI. Complete the Table

Pressure/Structure	Inspiration	Expiration
1. diaphragm	contract	relax
2. lung capacity	increase size	decrease size
3. thoracic capacity	decrease	increase
4. pleural pressure	decrease	increase
5. alveolar pressure	increase	decrease
6. transpulmonic pressure	increase	decrease

CHAPTER 21

I. Matching

1. pulmonary artery
2. pulmonary edema
3. angiotensin I
4. hypoxia
5. hypoxemia
6. pulmonary hypertension
7. bicarbonate ion
8. carbaminohemoglobin
9. embolus
10. foramen ovale

II. Dichotomous

1. artery
2. upper
3. less
4. increases
5. acidic, unloading
6. left
7. lower
8. higher
9. greater
10. passive, does not

III. True / False

True: 1, 2, 4, 7, and 8
False: 3, 5, 6, 9, and 10

IV. Sentence Completion

1. resistance to blood flow, increased cardiac output
2. 100 mm Hg
3. 16 mL O2/100 mL
4. 10 vol %
5. loading
6. left
7. aorta, pulmonary
8. pons, pulmonary stretch receptors
9. medulla, [H+] in cerebral spinal fluid
10. alveolar ventilation

V. Multiple Choice

1.c 2.b 3.a 4.c 5.a 6.e 7.e 8.a 9.e 10.c

VI. Complete the Table

Shift right:	1, 2, 3, and 5
Shift left:	4
Loading:	4
Unloading:	1, 2, 3, and 5

VII. Complete the Table

Increases:	1, 2, 3, and 5
Decreases:	4 and 6

CHAPTER 22

I. Matching

1. muscosa
2. portal circulation
3. mucin
4. short
5. long
6. gastrin
7. secretin
8. lower esophageal sphincter
9. peristalsis
10. pyloric sphincter
11. chief cells
12. acini
13. lipase, bile salts
14. microvilli
15. small intestine

II. Dichotomous

1. splanchnic
2. Short
3. gastrin
4. gastric
5. chief, stomach
6. direct
7. hyperosmotic
8. ileocecal
9. chylomicron
10. gastrocolic

III. True / False

True: 3, 4, and 10
False: 1, 2, 5, 6, 7, 8, and 9

V. Sentence Completion

1. vagus
2. parasympathetic, sympathetic
3. gastrin
4. Carbonic anhydrase
5. trypsin, enterokinase
6. Secretin
7. microvilli
8. Bile salts, pancreatic lipase
9. sodium
10. skeletal, voluntary

V. Multiple Choice

1.a 2.e 3.c 4.d 5.e 6.c 7.a 8.a 9.a 10.c

CHAPTER 23

I. Matching

1. glomerulus
2. proximal convoluted tubule
3. ureter
4. juxtamedullary nephron
5. luminal membrane
6. peritubular capillary
7. juxtaglomerular apparatus
8. glomerular hydrostatic pressure
9. fenestra
10. tubular transport maximum (Tm)
11. loops of Henle
12. vasa recta
13. colloid osmotic pressure
14. ADH
15. urea

II. Dichotomous

1. juxtamedullary
2. right, left
3. constrict, increasing, decreasing
4. greater, greater
5. increased
6. endocytosis, proximal
7. basolateral, luminal

III. True / False

True: 4, 7, 8, 9, and 10
False: 1, 2, 3, 5, and 6

IV. Sentence Completion

1. cortical, cortex
2. secreted
3. glucose, proteins
4. 4
5. 18 mm Hg
6. sodium transport
7. glucose threshold, Tm_G
8. distal
9. collecting ducts, sodium, water
10. ureters, bladder, urethra

V. Multiple Choice

1.a 2.c 3.c 4.b 5.d 6.b 7.c 8.e 9.e 10.e

VI. Complete the Table

None: 1, 4, and 5
Decrease: 2, 3, 6, and 7
Increase: 8

VII. Complete the Table (Answers in order: glucose, sodium, water)

1. decrease, none, increase
2. none, increase, none
3. none, increase, none
4. none, none, increase
5. none, none, decrease
6. decrease, decrease, decrease
7. none, decrease, decrease

CHAPTER 24

I. Matching

1. positive balance
2. osmoreceptor
3. baroreceptor
4. hyponatremia
5. ADH
6. micturition
7. urethra
8. aldosterone
9. atrial natriuretic factor
10. diuretic
11. hypokalemia
12. hypertension
13. azotemia
14. hemodialysis
15. detrusor

II. Dichotomous

1. extracellular, intracellular
2. sodium, potassium
3. input>output
4. decreases, increases
5. collecting duct
6. increase
7. decrease
8. increase, decrease
9. hypokalemia
10. peritoneal

III. True / False

True: 2, 5, and 7
False: 1, 3, 4, 6, 8, 9, and 10

IV. Sentence Completion

1. 2.5
2. A, B, less than
3. 20, 7.5, 2.5, 30, 10
4. hypothalamus, ADH
5. kidney, CV system
6. renin, granular cells of the juxtaglomerular apparatus
7. excretion
8. potassium, sodium, potassium, sodium
9. potassium, sodium
10. decreased

V. Multiple Choice

1.c 2.d 3.b 4.b 5.e 6.c 7.c 8.e 9.e 10.b

VI. Complete the Table

1. increase, increase
2. decrease, increase
3. increase, decrease
4. increase, increase
5. decrease, decrease
6. increase, increase
7. increase, increase
8. increase, increase
9. increase, none
10. increase, increase

VII. Complete the Table

1. increase
2. increase
3. increase
4. increase
5. decrease
6. decrease
7. increase
8. increase
9. increase
10. decrease
11. decrease
12. decrease
13. decrease
14. none

CHAPTER 25

I. Matching

1. acid
2. base
3. amphoteric
4. dissociation
5. association
6. buffer
7. chemorecepter
8. respiratory acidosis
9. respiratory alkalosis
10. metabolic acidosis

II. Dichotomous

1. acid, base
2. carboxylate, accept
3. proteins
4. dehydration
5. greater, lowered
6. inversely, basic
7. lower, lower
8. NH_3
9. increased
10. greater

III. True / False

True: 6 and 8
False: 1, 2, 3, 4, 5, 7, 9, and 10

IV. Sentence Completion

1. low
2. 3.9
3. 0.7-3.8
4. carbonic anhydrase
5. weak acid, conjugate base
6. left
7. peripheral, central
8. H_2CO_3, tubular epithelium
9. two
10. hydrogen, Na^+/H^+

V. Multiple Choice

1.e 2.d 3.e 4.a 5.c 6.a 7.c 8.c 9.a 10.c

VI. Complete the Table

Decrease: 1, 4, 6, and 8
Increase: 2, 3, 5, and 7

CHAPTER 26

I. Matching

1. filtered calcium
2. tetany
3. hydroxyapatite crystals
4. osteoid
5. osteoblast
6. osteoclasts
7. epiphyseal plate
8. chondrocyte
9. PTH
10. calcitonin
11. 1,25-dihydroxy-vitaminD_3
12. proteoglycan, collagen
13. phosphaturia
14. osteoporosis
15. rickets, osteomalacia

II. Dichotomous

1. Phosphorus
2. ionized
3. active transport
4. proximal, active transport
5. increase, decrease
6. osteoblasts
7. increase, decrease
8. resorption, osteoclasts
9. greater
10. greater

III. True / False

True: 1, 3, and 6
False: 2, 4, 5, 7, 8, 9, and 10

IV. Sentence Completion

1. Calcium, phosphorus
2. conjugate base HPO_4^{2-}, weak acid $H_2PO_4^-$
3. calcium, phosphorus crystals
4. collagen, proteoglycans
5. proximal tubule cells of kidney
6. Vitamin D_3^- cholecalciferol
7. osteoclasts
8. osteoid, hydroxyapatite
9. lower
10. chondrocytes, epiphysealplate

V. Multiple Choice

1.d 2.c 3.a 4.a 5.c 6.c 7.d 8.b 9.b 10.e

VI. Complete the Table

1. increase, increase, decrease
2. all decrease
3. all increase
4. decrease, decrease, increase
5. all decrease

CHAPTER 27

I. Matching

1. poikilotherm
2. rectal
3. mean skin temperature
4. thermography
5. mean body temperature
6. metabolic rate
7. thermal homeostasis
8. human calorimeter
9. respirometer
10. radiation
11. evaporation
12. convection
13. antirise response
14. heat exhaustion
15. hypothermia

II. Dichotomous

1. Homeotherms
2. less
3. more, radiation
4. constricted, decrease, radiation
5. greater
6. Rectal
7. core shell
8. before
9. Central
10. hyperthermia

III. True / False

True: 1, 2, 3, 6, and 10
False: 4, 5, 7, 8, and 9

IV. Sentence Completion

1. Cal/m^2/hr
2. lower
3. conduction, greater
4. colder, warmer
5. hottest
6. 2/3, 1/3
7. negative feedback
8. Antirise, hypothalamus, effector organ system
9. regulated, unregulated
10. interleukin I, cytokines, prostaglandin E

V. Multiple Choice

1.c 2.c 3.a 4.d 5.e 6.c 7.a 8.a 9.c 10.b

VI. Complete the Table

Increase: 1, 3, 4, 5, 7, and 9
Decrease: 2, 6, and 8

CHAPTER 28

I. Matching

1. lymph node
2. thymus gland
3. pathogen
4. antigen
5. antibody, immunoglobulin
6. anaphylaxis
7. B lymphocyte
8. cell-mediated immune response
9. helper T-cells
10. interleukins
11. lymphokines
12. MHC
13. menoclonal antibody
14. interferons
15. variable

II. Dichotomous

1. less, more
2. Afferent, efferent
3. greater
4. primary, secondary
5. nonspecific
6. I
7. B–
8. IgA
9. active
10. T_H

III. True / False

True: 2, 7, and 10
False: 1, 3, 4, 5, 6, 8, and 9

IV. Sentence Completion

1. subclavian veins
2. bone marrow, liver, yolk sac
3. histamine
4. cytokines
5. cell-mediated immunity
6. thymus
7. antigen presenting macrophage CD4+
8. Interleukin-2
9. Tc, Cytotoxic CD8+
10. two, variable

V. Multiple Choice

1.c 2.e 3.d 4.e 5.e 6.a 7.d 8.c 9.b 10.d

CHAPTER 29

I. Matching

1. polycythemia
2. acclimation
3. acclimatization
4. hypoxia
5. hyperventilation
6. carboxyhemoglobin
7. heat stroke
8. pneumoconiosis
9. atelectosis
10. anoxia
11. hypoxemia
12. free radical

II. Dichotomous

1. Adaption
2. arterial Po_2
3. acclimatization
4. increases
5. greater
6. lower
7. head to toe, positive
8. heat
9. Cadmium
10. respiratory

III. True / False

True: 5, 6, 7, 8, and 9
False: 1, 2, 3, 4, and 10

IV. Sentence Completion

1. superchiasmatic nucleus of the hypothalamus
2. bicarbonate
3. lungs, tissues, tissues, lungs
4. lungs, tissues, blood, lungs
5. 80
6. right
7. Non-shivering
8. ≤ 2, alveolar macrophages
9. paint
10. liver, fatty

V. Multiple Choice

1.c 2.e 3.c 4.c 5.d 6.b 7.e 8.b 9.c 10.e

CHAPTER 30

I. Matching

1. phosphagen energy system
2. static
3. $V0_2$ max
4. diffusing capacity of lung
5. dynamic
6. aerobic
7. recovey oxygen
8. hyperpnea

II. Dichotomous

1. Static
2. aerobic
3. anaerobically
4. lower
5. greater
6. high carbohydrate
7. III
8. hyperthermia
9. changes little
10. large, normal

III. True / False

True: 3, 4, 5, 6, and 10
False: 1, 2, 7, 8, and 9

IV. Sentence Completion

1. positive dynamic
2. O_2 mL/kg/min
3. fast
4. carbohydrate, lipid or mixed
5. ~25 fold, vasodilation
6. greater, lower
7. coronary area, skeletal muscle, brain, splanchnic, renal area
8. cardiovascular system
9. oxygen uptake
10. stroke volume, heart rate

V. Multiple Choice

1.a 2.c 3.b 4.b 5.a 6.a 7.e 8.c 9.a 10.a

VI. Complete the Table

Increase: 1, 3, and 5
Decrease: 2 and 4

CHAPTER 31

I. Matching

1. Leydig cells
2. corpus luteum
3. corpus luteum, granulosa cells
4. seminiferous tubules
5. LH
6. FSH
7. prostate
8. clitoris
9. secondary amenorrhea
10. blood-testis barrier
11. estrogen or GnRH
12. antrum
13. endometrium
14. epididymis
15. acrosome

II. Dichotomous

1. periphery
2. spermiogenesis
3. Smooth, sympathetic
4. LH, Leydig
5. One ovum
6. follicular
7. autocrine, paracrine
8. luteal phase
9. theca
10. granulosa

III. True / False

True: 1, 2, and 6
False: 3, 4, 5, 7, 8, 9, and 10

IV. Sentence Completion

1. four, one
2. 40
3. GnRH, hypothalamus
4. about 40
5. progesterone, androgens
6. 14
7. inhibit
8. corpus luteum, granulosa
9. three ova
10. clitoris

V. Multiple Choice

1.e 2.a 3.a 4.d 5.a 6.a 7.a 8.c 9.e 10.a

VI. Complete the Table

1. none, decrease
2. stops, none
3. stops, stops
4. none, stops
5. none, stops
6. stops, none
7. continues, continues
8. none, stops
9. none, stops
10. none, none

CHAPTER 32

I. Matching

1. cortical granule
2. endometrium
3. ectopic pregnancy
4. lactalbumin
5. oxytocin
6. amnion
7. capacitation
8. fallopian tube
9. morning sickness
10. acrosome
11. monozygous twins
12. embryoblasts
13. trophoblasts
14. placenta
15. hCG

II. Dichotomous

1. greater
2. epididymis
3. zygote
4. Monozygous
5. trophoblast
6. veins
7. hCS
8. second
9. embryonic
10. Prostaglandins

III. True / False

True: none
False: 1–10

IV. Sentence Completion

1. fallopian tube
2. progesterone
3. zona pellucida
4. calcium
5. trophoblast/chorion, uterine endometrium
6. placenta, corpus luteum
7. adrenal gland, placenta
8. Relaxin, Progesterone
9. eighth
10. posterior, oxytocin

V. Multiple Choice

1.a 2.d 3.c 4.c 5.d 6.e 7.a 8.c 9.d 10.b

CHAPTER 33

I. Matching

1. XX
2. genital tubercle
3. H-Y antigen
4. Müllerian duct system
5. Wolffian duct system
6. labioscrotal swellings
7. dihydrotestosterone
8. Klinefelter syndrome
9. adrenogenital syndrome
10. testicular feminization
11. excitement
12. orgasm
13. atrophic vaginitis
14. vaginismus
15. vasectomy

II. Dichotomous

1. sperm
2. Müllerian
3. earlier
4. does not require
5. female
6. Klinefelter syndrome
7. vasocongestion
8. orgasm
9. efferent
10. infertility

III. True / False

True: 4, 6, and 8
False: 1, 2, 3, 5, 7, 9, and 10

IV. Sentence Completion

1. y chromosome
2. Sertoli, Leydig
3. Turner
4. clitoris pituitary
5. plateau
6. estrogen
7. orgasm
8. psychological
9. FSH, LH, anterior
10. estrogen, implantation

V. Multiple Choice

1.e 2.a 3.c 4.d 5.e 6.a 7.a 8.e 9.a 10.c